AF601875

ABOUT THE AUTHORS

Vikram Singh has completed his doctorate degree from Himachal Pradesh University, Shimla. His specialisation is biodiversity especially faunal diversity. His areas of interest are the issues of biodiversity, wild life ecology, wildlife conservation, sustainable development and environment. He has actively worked on issue of wild animal-human conflict especially man monkey conflict in Himachal Pradesh. Presently he is working as Assistant Professor in Zoology (ad hoc) in Department of Zoology, Acharya Narendra Dev College Govindpuri, Delhi. He is also working on the status and ecology of *Semnopthicus ajax* in Chamba. Previously he has also worked on antiplasmodial effects of plant extracts against malaria parasite.

Professor H.S. Banyal, a renowned protozoologist, has dedicated decades for studies on malaria. He was Professor, Chairperson and Dean of Department of Biosciences, Himachal Pradesh University, Shimla-5 and Professor & Chairman. Department of Zoology, Panjab University, Chandigarh, India. He also remained DSW and Dean of Studies in Himachal Pradesh University, Shimla. Prof. Banyal specializes in immunology and biochemistry of malaria parasites and other blood protozoans. Also he is keen in understanding the biology and biodiversity of wild fauna of Himachal Pradesh. Presently he is the Vice Chancellor of Abhilashi University, Mandi, Himachal Pradesh.

Faunal Diversity of Khajjiar Lake, District Chamba, Himachal Pradesh

Faunal Diversity of Khajjiar Lake, District Chamba, Himachal Pradesh

–Authors–
Vikram Singh
H.S. Banyal

2015
Scholars World
A Division of
Astral International Pvt. Ltd.
New Delhi - 110 002

© 2015 AUTHORS
ISBN *9789390384198* (Hardbound)

Publisher's note:
Every possible effort has been made to ensure that the information contained in this book is accurate at the time of going to press, and the publisher and author cannot accept responsibility for any errors or omissions, however caused. No responsibility for loss or damage occasioned to any person acting, or refraining from action, as a result of the material in this publication can be accepted by the editor, the publisher or the author. The Publisher is not associated with any product or vendor mentioned in the book. The contents of this work are intended to further general scientific research, understanding and discussion only. Readers should consult with a specialist where appropriate.
Every effort has been made to trace the owners of copyright material used in this book, if any. The author and the publisher will be grateful for any omission brought to their notice for acknowledgement in the future editions of the book.
All Rights reserved under International Copyright Conventions. No part of this publication may be reproduced, stored in a retrieval system, or transmitted in any form or by any means, electronic, mechanical, photocopying, recording or otherwise without the prior written consent of the publisher and the copyright owner.

Published by **Scholars World**
A Division of
Astral International Pvt. Ltd.
– ISO 9001:2008 Certified Company –
4760-61/23, Ansari Road, Darya Ganj
New Delhi-110 002
Ph. 011-4354 9197, 2327 8134
E-mail: info@astralint.com
Website: www.astralint.com

Laser Typesetting : **SSMG Computer Graphics**, Delhi - 110 084

Printed at : **Thomson Press India Limited**

Preface

Biodiversity, that is, plants, animals and other organisms, contribute to the provision of numerous ecosystem services, such as medicinal, agricultural products and clean water.

Biodiversity, encompassing variety and variability of all life on earth, is the product of over 3.5 billion years of evolutionary history. It has been generally defined as the *'full variety of life on Earth'*. Human society relies on biological resources and their diversity, and the ecosystems that sustain them, to provide essential goods and services.

United Nations General Assembly has declared 2011-2020 as "United Nations Decade of Biodiversity". The strategic goals incorporated under the declaration include, preventing the extinction of endangered species, halving the loss of forests and natural habitats, sustainable fishing, halting ocean fertilisation, saving coral reefs, reducing pollution, increasing protected terrestrial and inland water areas from 12 to 17% and increasing protection for marine and coastal areas from 1to 10%. If the present rate of decline continues, half of the world's species will get extinct in 21st century. It has been predicted that 20% of the world's species would get extinct within next 30 years and at least 50% in the decades that follow. Presently biodiversity on the earth is being impoverished at an alarming rate, just at the time when mankind needs it most for sustaining its own life. It is now well recognised that the well-being of human beings and biodiversity are more interdependent than ever before. Virtually all governments, organisations and communities have responded to this situation in several ways.

The Himalayan ecosystem is unique and perhaps no other single geographical feature had greater influence on the life, culture and history of the people of Indian subcontinent than these mountains. In recent years, the state of Himachal Pradesh has come under a strong threshold of development. Natural ecosystems/habitats have been over-exploited and even destroyed by the rapidly increasing human population and tourist inflow. A number of endemic and restricted range species found in the area/region are facing threat to their existence.

The bio-diversity in Khajjiar lake area, like other parts of Himachal Pradesh is very rich and diversified, but, in recent years, areas of Khajjiar in particular and the entire state in general have come under a strong threshold of development. Natural ecosystems/habitats have been over-exploited and even destroyed by the rapidly increasing human population and tourist inflow. A number of endemic and restricted range species found in the area/region are facing threat to their existence. The area Khajjiar sanctuaryis one of the oldest conservation areas for wildlifein Himachal Pradesh and, being a favoured touristdestination, is also under remarkable anthropologicalpressure which may severely influence habitatconservation.

Present books provide the details of fauna ofKhajjiar along with their status, habitat, habits and other ecological informations. This book can be used as a reference to base line survey of the fauna of Khajjiar. An effort is also there to mention thethreats to biodiversity and to propose some of the conservation measures.

Authors

Contents

Abbreviations

%	Percent
&	And
°C	Degree Celsius
amsl	Above mean sea level
AqA	Aquatic animal
C	Common
CITES	Convention in trade of endangered species cm Centimetre
CR	Carnivorous
e.g.	For example
et. al.	Etalia; and others
etc.	Etcetera; and others
ex	Example
F.	Fabricius
F. H.	Feeding habits
Fig.	Figure
FR	frugivorous
GR	Graminivorous
H.P.	Himachal Pradesh
I	Insectivorous
i.e.	That is
L.	Linnaeus
m	Metre
mm	Millimetre

No.	Number
OC	Occasional
OM	Omnivorous
R	Resident
R/LM	Resident with local movements
R/SV	Resident with summer influx
R/WV	Resident with winter influx
Ra	Rare
RA	Relative abundance
RS	Residential status
SC	Scavengers
sp.	Species (singular)
spp.	Species (plural)
Sq. km	Square kilometre
Sq. m	Square metre
SV	Summer visitor
UC	Uncommon
VC	Very common
VgM	Vegetable matter
viz.,	Videlicet, namely
WPA	Wildlife Protection Act
WV	Winter visitor

Chapter 1

Biodiversity

Biodiversity, encompassing variety and variability of all life on earth, is the product of over 3.5 billion years of evolutionary history. It has been generally defined as the *'full variety of life on Earth'*. More specifically, biodiversity is the study of the processes that create and maintain variations. It is concerned with the variety of individuals within populations, the diversity of species within communities, and the range of ecological roles within ecosystems. Biological diversity or biodiversity refers to the diversity of life.Biodiversity is the result of evolutionary plasticity of living organisms, and it increased geometrically through perhaps 3.5 billion years, proliferating by trial and error, controlled by natural selection, filling almost every one of the habitable ecological niches created in a likewise evolving world environment. The variability among living organisms from all sources including terrestrial, marine and other aquatic ecosystems, and the ecological complexes of which they are part; this includes diversity within species, between species, and of ecosystems.

For times immemorial, man has been fascinated by the diversity of life. Perhaps it has not attracted the same attention as global warming and ozone depletion but it has certainly been catapulted into the centre stage of the world-wide environmental policies in handful of years. Earth encompasses many characteristics of life, but none is as extraordinary as its sheer diversity, the seemingly limitless array of life-forms, the vast genetic variety they contain, the rich ecological associations they form, and the complex behaviours they adopt in meeting the challenge of survival. This diversity reflects life's capacity to occupy the ever-changing spectrum of environment on the planet, from arctic ice cap to dense tropical rainforest, from sun-bleached desert to sunless subterranean cavern, from mountain summit to ocean depth.

The word biodiversity which is the abbreviated word for biological diversity appears to have come into prominence around 1980, when Norse and McManus (1980) first defined it. Its abbreviation into 'biodiversity' was apparently made by Walter G. Rosen in 1985 during the first planning meeting of the 'National Forum on Biodiversity' held at Washington DC in September 1986 (UNEP, 1995). The book entitled *Biodiversity* (Wilson and Peters, 1988) introduced the notion of biodiversity and popularised this word among the scientific community as well as among the public. Since then, not only the number of publications on biodiversity increased, but the number of people interested in the subject for one reason or the other has also steadily increased.

The United Nations General Assembly has declared 2011-2020 as "United Nations Decade of Biodiversity". The strategic goals incorporated under the declaration include, preventing the extinction of endangered species, halving the loss of forests and natural habitats, sustainable fishing, curbing ocean fertilisation, saving coral reefs, reducing pollution, increasing protected terrestrial and inland water areas from 12 to 17% and increasing protection for marine and coastal areas from 1to 10%. Also included is a provision requiring governments to reclaim 15% of degraded lands, in an attempt to get nations to expand from conservation into restoration (Aitken, 2011). A land mark step in conservation of bio-resources was initiated on December 29, 1993 when 'Convention on Biological Diversity' came into force with the objectives of conservation of biological diversity, sustainable use of the components of biological diversity and equitable sharing of the benefits arising out of the utilization of its genetic resources. UNEP, UNESCO, UNDP, CITES, Centre for Biodiversity Conservation, Conservation International and many other NGOs have come forward with Secretariat of the Convention on Biological Diversity (CBD) to achieve conserve biodiversity and 22nd May of every year is celebrated as the International Day for Biodiversity.

Variability among living things is generally considered at three levels i.e. genetic, species and ecosystem level. Genetic diversity refers to variability in the genes that every individual inherits from its parents and passes on to the next generation. It is the sum total of varied genetic information contained in the genes of individual plants, animals and micro-organisms that inhabit the earth. Diversity within a species enables it to adapt or resist changes in environment, climate and agricultural methods or in the presence of new pests. This represents variations in genes and is due to the variations at the molecular level involving the sequences of bases in nucleic acids, which constitute the genetic code (Melchias, 2001; Kumar and Asija, 2006).

Species' diversity refers to the variety of living organisms (wild or domestic) within a geographic area. Species' diversity is often measured with the goal of examining and comparing patterns of species' distribution at local and regional levels. These measures are used chiefly for examining particular groups of species rather than the whole range of species and interactions found in nature. Complex spatialpatterns of species' diversity have often been recognised by dividing species richness into three major components to characterise diversity on different scales. Thus, the species richness is considered as Alpha-diversity, Beta-diversity and Gamma-diversity. Alpha-diversity of species is described as species richness within a community or habitat. Beta-diversity is a measure of the rate and extent of change in species along a gradient from one habitat to others. Gamma-diversity is the richness of species of a range of habitat in a geographical area which is a consequence of alpha-diversity of the habitats, together with the extent of beta-diversity between them (Krishnamurthy, 2003; Belsare, 2007).

Ecosystem diversity is the intricate network of different species present in local ecosystems and the dynamic interplay among them. Measuring ecosystem diversity is difficult because each of the Earth's ecosystems merges into the ecosystems around it. Ecosystem consists of interdependent communities of species and their physical environment. The extent of an ecosystem or habitat is imprecise; a single ecosystem may be as large as thousands of hectares or as small as a drop of water. They include major natural systems such as grasslands, mangroves, coral reefs, wetlands and tropical forests as well as agricultural ecosystems that, while depending upon human activity for their existence and maintenance, have characteristic assemblage of plants and animals (Ananthakrishnan and Sivaramakrishnan, 2006).

Values attached to biodiversity can be classified into three categories: Productive Use Value, Consumptive Use Value and Indirect Value. Productive Use Valueis a value assigned to products that are commercially harvested for exchange in formal markets and is, therefore, the only value of biological resources that is reflected in national income accounts. Products such as fuel wood, timber, fish, animal skins, musk, fodder, fruits, cereals and medicinal plants come under this category. On the other hand, Consumptive Use Value is the value placed on natural products that are consumed directly. The value of such goods can be considerable. A significant number of such non-timber forest products as soft broom grass and cane fall under this category. Indirect values are related primarily with the functioning of ecosystems, do not normally appear in national accounting systems, but they may far outweigh consumptive and non-consumptive values. Maintenance of ecological balance and prevention of soil erosion are examples of such indirect values.

Biodiversity is our natural wealth. Its conservation is important for both economic and ethical reasons. It provides us goods and services fundamental to our survival, including clean air, fresh water, medicines and shelter. It enables us to adapt to changing needs and circumstances.

For example, forested ecosystems provide us fuel, medicine, construction material and wildlife habitat; wetlands and riparian areas protect water quality and aquatic life; oceans provide food and regulate climate; and agro-ecosystems produce food. Biodiversity also provides people with recreational, psychological, emotional and spiritual enjoyment. Some people opine that we should protect and restore biodiversity because of its benefits to mankind, while others believe that it is our moral obligation to care about biodiversity simply because all species have the right to live and to be valued in nature, whether we understand their benefits to human world or not (Alonso *et al.*, 2001).

India is situated north of the equator between 66°E to 98°E and 8°N to 36°N. It is bordered by Nepal, China and Bhutan in the north; Bangladesh and Myanmar in the east; the Bay of Bengal in the south east; the Indian Ocean in the south; the Arabian Sea in the west; and Pakistan in the north-west.

The varied edaphic, climatic and topographic conditions have resulted in a wide range of ecosystems and habitats such as forests, grasslands, wetlands, coastal, marine and deserts ecosystems. India represents: (i) Two 'Realms'- the Himalayan region represented by Palearctic Realm and the rest of the sub-continent represented by Malayan Realm; (ii) Five Biomes e.g. Tropical Humid Forests; Tropical Dry Deciduous Forests (including Monsoon Forests); Warm Deserts and Semi-Deserts; Coniferous Forests; Alpine Meadows; and (iii) Ten biogeographic zones and Twenty-seven biogeographic provinces.

India is situated at the tri-junction of the Afro-tropical, the Indo-Malayan and the Paleo-Arctic realms, which display significant biodiversity. Being one of the 17 identified megadiverse countries, it is home to 8.58% of mammalians, 13.66% of avians, 7.91% of reptilians, 4.66% of amphibians, 11.72% of fish, and 11.80% of plant species documented so far. From the biodiversity standpoint, India has some 59,353 insect species, 2,546 fish species, 240 amphibian species, 460 reptile species, 1,232 bird species and 397 mammal species, of which 18.4% are endemic and 10.8% are threatened. India's forest cover ranges from the tropical rainforest of the Andaman Islands, Western Ghats, and North-Eastern India to the coniferous forest of the Himalayas. Between these extremes lie the Sal-dominated moist deciduous forests of Eastern India, the Teak-dominated dry deciduous forests of Central and Southern India, and the Babul-dominated thorn forests of Central Deccan and Western Gangetic plains. Among species found in India, only 12.6% of mammals and 4.5% of birds are endemic, as against 45.8% of reptiles and 55.8% of amphibians. India contains 172 (2.9%) of the IUCN designated threatened species. It has been estimated that at least 10% of the country's recorded wild flora, and possibly the same percentage of its wild fauna, are on the threatened list, many of them on the verge of extinction (Varughese *et al.*, 2009).

Traditional and substantial dependence on biodiversity and avifaunal resources for fodder, fuel wood, timber and minor forest produce has been an accepted way of life for the rural population that accounts for nearly 74% of India's population. With radical demographic changes, the land to man ratio and forest to man ratio has rapidly declined. The lifestyles and the biomass resource needs having remained unchanged, the remnant forests have come under relentless pressure of encroachment for cultivation, and unsustainable resource extraction, rendering the very resource base unproductive and depleted of its biodiversity. Today this diversity of life is threatened by human activities, although the exact rate of species loss is difficult to ascertain. These activites are unabated human population growth, overexploitation of resources, pollution, and global climatic change. Disappearance of species is not an aberrant process in the course of time. Biologists have estimated that the natural rate of extinction is about one per one million species in a year which is also referred as 'background' rate of extinction. In the deep past, species were also wiped out in large scales due to extrinsic factors that were beyond normal environment regime. Species' evolution and extinction are very much part of evolutionary history of biotic world. But the concern is the alarming rate at which species are becoming extinct today due to reckless alteration and degradation of environment quality and putting the very future in jeopardy. The current rate of species' extinction is about 1000 times faster, while the evolution of new species is limited by evolutionary constrains (Saikia *et al.*, 2010).

As per the IUCN Red List (2008), India has 413 globally threatened faunal species, which is approximately 4.9% of the world's total number of threatened faunal species.

These include 53 species of mammal, 69 birds, 23 reptiles and 3 amphibians (Varughese *et al.*, 2009). India contains globally important populations of some of Asia's rarest animals, such as Asiatic Lion, the Bengal Tiger, and the Indian White-Rumped Vulture. The Indian White-Rumped Vulture was on the verge of extinction due to feeding on the carrion of diclofenac-treated cattle. At least, 10% of the country's recorded wild flora and possibly the same percentage of its wild fauna are estimated to be in the threatened list while many of them are on the brink of extinction.

If the present rate of decline continues, half of the world's species will get extinct in 21st century. It has been predicted that 20% of the world's species would become extinct within next 30 years and at least 50% in the decades that follow. Presently biodiversity on the earth is being impoverished at an alarming rate, just at the time when man needs it most for sustaining his own life. It is now well recognised that the well-being of human beings and biodiversity are more interdependent than ever before. Virtually all governments, organisations and communities have responded to this situation in several ways.

The Himalayan ecosystem is unique and perhaps no other single geographical feature had greater influence on the life, culture and history of the people of Indian subcontinent than these mountains. In addition, these are extremely rich repositories of natural resources and biological wealth (Jairajpuri, 1993). Himalayas are the most magnificent complex folded and the youngest mountain systems in the world and form a physical barrier between the high plateaus of Tibet and Central Asia, and the Indian plains extending from river Indus in the west to the river Brahmaputra in the east. The mountain ranges are approximately 2500 Kms long with a total width varying between 40-400 Kms. It is situated between latitudes 27° and 36° North and longitudes 70° and 96° East. Its location and great expansion in latitude and altitude offers a wide variety of habitats each supporting its own distinctive type of fauna (Mehta and Julka, 2002). Geographically, Himalayas are divisible into Eastern Himalayas, the Central or Nepal Himalayas and Western Himalayas which includes the State of Himachal Pradesh.

Himachal Pradesh, mainly a hilly state, lies in the northwest Himalayas between 30° 22' to 33°12' North latitude and 75° 47' to 79° 04' East. The physiography of Himachal Pradesh is almost mountainous with elevations ranging from 350 to 6975 metres above mean sea level and total area of the state is 55,673 sq km. Its northern border is bounded by Tibet, whereas, in the northwest, it has a common border with Kashmir and the eastern border of the state is common with the hills of Uttarakhand, Haryana falls on the south and Punjab on the southwest (Mani, 1981; Chauhan, 1998).

Himachal Pradesh is divided by general increase in elevation from west to east and from south to north into following four biogeographical regions viz., Shivalik or Outer Himalayas, Lower or Lesser Himalayas, Higher or Greater Himalayas, and Trans Himalayas covering around 10.54% of Himalayan land mass. The Shivalik ranges are the southernmost zone of about 40 to 60 km width, comprising several highly eroded low ridges which covers the districts of Sirmour, Solan, Bilaspur, Hamirpur, Una and parts of Chamba and Kangra. The lesser Himalaya (about 80 km wide) runs from North of the Shivalik and parallel to the great Himalayan range this zone covers districts of Shimla, Mandi and parts of districts Chamba, Kullu, Kangra and Sirmour. The Great Himalayan ranges lie just north of the Chandrabhaga river in Lahaul-Spiti and Pangi region of Himachal Pradesh. This range is nearly 24 km wide and comprises the Great peaks rising up to an elevation of over 6000 m amsl. This zone covers Pangi region of Chamba district and certain portions of Kullu and Kinnaur districts. Varied physiographic and climatic factors have given rise to diverse natural ecosystems/habitats, namely, forests, grasslands and pastures, river, lake and wetlands, glaciers etc. in this region (Mehta and Julka, 2002). The Trans-Himalayan region comprises some parts of Lahaul and Spiti and Kinnaur district characterized by extreme cold, low precipitation and lack of vegetation and is often referred to as the cold desert (Rodger and Panwar, 1988).

Average rainfall in the state stands at 1523 mm, although it varies from a minimum of 300 mm at Lahaul and Spiti to a maximum of 4400 mm at Dharamsala. The temperature in the state varies according to elevation. From the end of February, mercury rises gradually till June, which is generally the hottest month in this region. With the onset of monsoons, there is a gradual fall in temperature. When the monsoon ends by middle of September, temperature falls gradually in the beginning and fairly rapidly after November (Mani, 1981; Vedwan and Rhodes, 2001; Thakur, 2008).

On the basis of agro-climatic condition, Himachal Pradesh can be broadly classified into four major zones as subtropical (low lying hills), subtemperate (mid hills), temperate (high hills and interior valleys) and cold and dry zones.

a) **Subtropical Zone:** This zone comprises of valleys and low lying hills near the plains of Punjab and Haryana. Altitude in this zone varies from 350 to 1000 meters with an annual rainfall between 600 to 1000 mm. This zone is very fertile and subjected to intensive cultivation.

b) **Sub temperate Zone:** This zone comprises of mid hills and the altitude varies from 1000 to 1500 meters above mean sea level. The climate of this zone is moderate with annual precipitation ranging from 900 to 1000 mm.

c) **Temperate High Zone:** This zone consist of of high hills and interior valley areas with altitudes varying from 1550 to 3000 meters above mean sea level. Annual rainfall varies from 900 to 1000 mm and snowing in winter is a usual feature.

d) **Cold and Dry Zone:** This zone lies between 3000 to 3700 meters above mean sea level. Annual rainfall is scanty and varies from 250 to 400 mm. This zone is extremely cold and minimum temperature on an average comes down to 15°C.

Natural vegetation of the state can be classified into six broad types of forests viz., Tropical forests (confined to foothills), represented by two subtypes, namely thorn-scrub of *Acacia* and *Zizyphus,* and dry deciduous forests of *Shorea robusta;* Subtropical forests (500-1800 m), which are further composed of two subtypes i.e. subtropical dry evergreen forests of *Terminalia, Albizzia, Olea* etc (below 1200 m), and subtropical pine forests of chir pine (*Pinus roxburghii*) found upto 1800 m; Temperate forests (1500-3000 m) which are also divided into two subtypes i.e. Himalayan moist temperate covers areas between 1500 and 3000 m, where the flora is dominated by oaks (*Quercus* spp.), deodar (*Cedrus deodara*), fir (*Abies pindrow*), blue pine (*Pinus wallichiana*) and horse chestnut (*Aesculus indica*); and another is the Himalayan dry temperate subtype of Holm oak (*Quercus ilex*) and edible pine (*Pinus gerardiana*) which is best developed at 2000 to 3000 m in the greater Himalayan regions of upper Sutlej valley in Kinnaur district; and Sub-Alpine and Alpine vegetation (above 4000 m) which is dominated by Birch and Rhododendron and interspersed with high-altitude meadows, found in most parts of Lahaul, Spiti and Kinnaur districts of the state (Mehta, 2005; Narwade *et al.*, 2006; Thakur, 2008).

The faunal and floral diversity in Himachal Pradesh is also very rich and diversified, primarily due to varied climatic conditions ranging from tropical in the foothills to arctic environment in the Trans-Himalayan region. Moreover, historical influx of fauna from adjacent biogeographical regions and subsequent speciation in relation to local environment has greatly enriched the animal resources of the area. There is a pronounced dominance of Palaearctic and endemic animals above timber line (3000 m), and largely Oriental and some Palaearctic and some Ethiopian elements at lower and middle altitudes. Thus rich biodiversity of Himachal Pradesh has sustained population and hill communities from times immemorial. But, in recent years, the state has come under a strong threshold of development. Natural ecosystems have been over-exploited and even destroyed by the rapidly increasing human population. A number of endemic and restricted range species found in the region are facing threat to their existence.

Rich diversity of animals in Himachal Pradesh is reflected by the presence of 2,542 faunal species belonging to different groups as compared to 89,500 animal species of the country (Mehta, 2005). Himachal Pradesh has a small geographical area of 55,673 square kilometres which is only 1.7% of India but it harbours more than 7% of the total fauna of the country. Invertebrates constitute 88.4% and vertebrates 11.6% of the fauna in Himachal Pradesh. Insects and other arthropods form a predominant group (4,641 species) among invertebrates, whereas vertebrates are dominated by birds comprising about 447 (610, revised) species (Mehta, 2005).

In Himachal Pradesh. Insects and other arthropods form a predominant group (4,641 species) among invertebrates, whereas vertebrates are dominated by birds comprising about 447 (610, revised) species (Mehta, 2005).

Chapter 2

Status of Biodiversity

Diversity is a property and not an entity in itself. It refers to the property of a set of objects of not being identical, of varying from each other in one or more characteristics. When applied to organisms, it refers to the universal attribute of all living things that each living individual is unique, that is, no two organisms are identical. The origin of this variability is to be found in the basic and fundamental property of the nucleic acids' molecules that the order of basis does affect the free energy of the molecule, in other words all combination of four bases that form the genetic code are chemically equally viable. This characteristic combined with natural selection allows the acquisition and accumulation of favourable mutations, and in approximately three thousand million years that life has existed on earth these processes have produced enormous biological variation that we see today, which is at best only a very small percentage of all the variation that has existed in the past (Solbrig, 2000).

The term biodiversity describes the number, variety and variability of living organisms. It refers to variability within living world. Ecosystem, genus and species are terms used to define it. Its fundamental discipline is systematic which includes classification, identification and nomenclature. It was Whittaker(1972) who introduced the concepts of alpha, beta and gamma diversity. The global distribution of species richness increases with decreasing latitude which means that total number of species per unit area increases in ascending order in polar, temperate and tropical region. In terrestrial ecosystems with increasing altitude the diversity generally decreases, meaning thereby thatthere is low species diversity in high altitude areas. The diversity in mid altitude may be higher than in lower altitude specifically in tropical forests but there is lack of substantial data and it has been observed in desert mountains of Arizona where in higher altitude and lower altitude, diversity is believed to be affected by lower temperature and aridity respectively. The precipitation too, is an important factor, as it increases; the terrestrial diversity also increases.

Biodiversity benefits human societies in a myriad of ways by providing a wide range of ecological, economic, social, cultural, educational, scientific and aesthetic services to our life supporting system (Goyal and Arora, 2009). Biodiversity, besides its ecological and intrinsic values, represents a considerable socioeconomic and monetary asset as well. Human society relies on biological resources and their diversity, and the ecosystems that sustain them, to provide essential goods and services (Noss, 1990). The greatest diversity of lifeforms was achieved during present geological period. Higher groups like vertebrates, flowering plants, etc., become most diverged only 30,000 years ago, very recently in geological terms. Earth is probably holding 10 million species of which approximately 1.7 million have been described by scientists (Saikia *et al.*, 2010 a).

A number of field biologists have carried out studies on different groups of biodiversity throughout the country. Out of total 86,874 animal species found in our country, insects alone comprise 68.32 percent while chordates represent only 5.70 percent (Alfred *et al.*, 1998). Of the vertebrates, around 1300 species of breeding, staging and wintering birds belonging to 88 families and 22 orders are known from India (Manakadan and Pittie, 2001). According to an estimate, there are about 89,451 described faunal species in India, although only less than half of its geographical area has been fully surveyed. Nemertinea, Pogonophora, Priapulida and Pentastomatida are invertebrate groups yet to be reported from India. Many other faunal groups are still to be documented in detail. The fauna of British India is available as baseline literature whose studies undertaken by various workers dates back to 1800. Zoological Survey India (ZSI) has published monographs and taxonomic revisions of many taxa together with 3929 new records during 1916-1991 (Chavan *et al.*, 2004).

Some biodiversity studies carried out in Himachal Pradesh revealed the rich faunal heritage of the state. Mehta (2005) has documented 2542 total animal species in varied ecosystems of the state of Himachal Pradesh. Major groups are 88 species of Odonata (Kumar, 2005), 167 species of scarabaeid beetles (Chandra, 2005), 288 species of Rhopalocera (Arora *et.al.*, 2005), 184 species of Geometrid moths (Walia, 2005), 319 species of Ichneumonid wasps (Jonathan, 2005), 19 species of Scolopendrid centipedes (Khanna, 2005), 104 species of Pisces (Mehta and Uniyal, 2005), 447 species of Aves (Mahabal, 2005) and 107 species of Mammalia (Chakraborty *et.al.*, 2005).

In the recent past, some studies have been undertaken in biodiversity-rich and important parts of the state, as a report on the faunal resources of Pin Valley National Park by Sharma (2008) elucidateds the presence of 183 species belonging to 147 genera of Oligochaeta, Mollusca, Crustacea, Scorpionida, Arachnida, Scolopendromorpha, Orthroptera, Dermaptera, Lepidoptera, Coleoptera, Diptera, Hymenoptera, Pisces, Herpetofauna, Aves and Mammalia. Singh (2007) has enumerated the presence of 31 species of insects belonging to five different orders viz. Orthroptera (seven species), Diptera (seven species), Lepidoptera (eight species), Coleoptera (three species) and Hymenoptera (six species) from Chandertal lake. Moreover, the distribution pattern of insects in different altitudes of Shimla hills has been studied by Thakur *et al.* (2008) who described 70 species of insects spread over four orders namely Coleoptera, Orthoptera, Lepidoptera and Hymenoptera.

Nematoda

Nematoda consists of mostly worm like invertebrates having appendage-less and un-segmented body. They vary in length from 82 ìm to over 8 m, and generally have a cylindrical body while a few may be fusiform, saccate or kidney shaped. There are some estimates that there may be as many as half a million more unknown species of roundworm yet to be discovered. The estimated number of species of nematodes from the world is around 500,000 though; there are some 30,000 described species of nematodes from the world. If the estimated number of species is anywhere close to correct, it would mean that roundworms are the second most diverse group of animals, trailing behind only the arthropods. Around 3,000 species have so far been reported from India. Of these more than 1200 are parasitic on vertebrates and invertebrates, and rest are plant parasitic and free-living (Baqri, 1998).

Nematodes have been divided into five groups on the basis of habitat and feeding habits, viz., vertebrate parasites, invertebrate parasites, plant parasites, microbotrophic, saprophagus or free-living and predacious. The nematode parasites of vertebrates, invertebrates and plants cause significant damage to their hosts, i.e. human beings, livestock and crops. They cause some serious diseases like onchocerciasis, ascariasis, filariasis etc. in humans. All the nematodes are not enemies. Free-living species are abundant, including nematodes that feed on bacteria, fungi and other nematodes. The nematode parasites of insects are considered as good biological control agents of pests. Others maintain natural balance in the soil. Recently, Rizvi (2010) has reported the presence of 20 genera of soil nematodes from Ladakh, of which 17 were new to the area and 14 were new to the state of Jammu and Kashmir.

Platyhelminthes

Plathelminthes (*platy*, meaning flat, *helminth*, meaning worm) are commonly known as flatworms having dorsoventrally flattened body. Animals of this phylum are relatively simple bilateral, unsegmented, soft-bodied invertebrate animals. Unlike other bilaterians, platyhelminthes have no internal body cavity and are, therefore, described as acoelomates.

Platyhelminthes are divided into four classes namely Turbellaria, Monogenea, Digenea and Cestoidea. It is highly variable group exhibiting ecto-and endo-parasitism. Turbellarians are mostly non- parasitic and free living, few are commensals. Rest of the three groups are entirely parasitic. Most of these cause various diseases among wild and domesticated animals including humans. Turbellarians are mostly predators, and live in water or in shaded, humid terrestrial environments. Flatworms mostly survive and propagate in tropical countries because they require optimum conditions for their survival. The distribution of parasitic forms is according to their respective hosts. The distribution of flatworms is confined to limited geographical area due to limited distribution of their intermediate hosts like molluscs.

About 25,000 species of Platyhelminthes are known from the world, of which about 4,000 species belong to class Turbellaria. In India, endemicity of Platyhelminthes has been reported in maximum number of species in Digenea (500 spp.), followed by Cestoda (350 spp.), Monogenea (250 spp.) and least in Turbellaria (60 spp.) (Ghosh, 1998).

Annelida

Annelids (*anellus* means little ring) are also called 'ringed worms'. Annelids are bilaterally symmetrical, triploblastic, coelomate worms with metameric segmentation. These worms play a pivotal role in agriculture and pharmaceutical industries. This phylum includes three classes namely Polychaeta, Oligochaeta and Hirudinea.Earthworms (Oligochaeta) are important because of their significant role in decomposition of surface litter, thus increasing the soil fertility by its redistribution and incorporation in the soil. Moreover, during unfavourable period, the microbial decomposition of dead worms releases considerable amount of nitrogen and other nutrients for growing vegetation.

There are over 17,000 living annelid species (Rouse, 2002), ranging in size from microscopic to the Australian Giant Gippsland earthworm, which can grow up to 3 metres (9.8 ft) long (Ruppert *et al.*, 2004). Out of this about 12,000 species belong to Polychaeta. This group is found almost throughout India. Indian Annelids have 3 classes, 80 families, 312 genera and 840 species, forming 6.6% of the global annelid fauna. Approximately 4,400 different species of earthworms have been identified worldwide (Sinha, 2009). There are 590 species of earthworms reported from India (Bisht *et al.* 2003; Tripathi and Bhardwaj 2004; Sathianarayanan and Khan 2006; Karmegam and Daniel 2007; Chaudhuri *et al.* 2008). Class Oligochaeta in India is represented by 381 species belonging to 87 genera and 14 families (Julka, 1998). Of these, 46 species have been reported from Himachal Pradesh (Paliwal, 1994; Julka and Paliwal, 1995; 2005). The taxonomic monographs on Indian Oligochaeta have been compiled by Stephenson (1923), Gates (1972) and Julka (1988).

Himachal Pradesh harbours subtropical and temperate Oligochaeta species. It was Gates (1972) who remarked that *Perionyx excavatus* and *Perionyx sansibaricus* found in Himachal Pradesh have been widely distributed within and outside India. Julka (1981) described *Anthropochorous lumbricids* from high altitude area of Lahaul valley. Paliwal (1994) reported 42 species from Himachal Pradesh including four species, *Plutellus sadhupulensis, Perionyx barotensis, Perionyx bainii* and *Perionyx simlaensis* as endemic to the state. Julka and Paliwal (2005) reported the existence of 46 species of earth worms from the state and of these only seven species are known from Lahaul & Spiti district. Paliwal (2008) described only one species (*Octolasioin tyrtaeum*) of Lumbricidae family from Pin Valley National Park in Lahaul & Spiti district. Recently, from neighbouring Kashmir valley, Najar and Khan (2011) has reported the presence of eight species of earthworms, belonging to 5 genera, three families (Moniligastridae, Megascolecidae and Lumbricidae) and two orders under class Oligochaeta.

Crustacea

Crustacea is a large group of phylum Arthropoda, sometimes treated as subphylum, which includes animals like crabs, lobsters, crayfish, shrimp, krill and barnacles. These arthropod species are found mostly in marines,and little fresh water sourcesbut while some are terrestrial, others are parasitic (like tongue worm) and some are sessile (like barnacles). Body of these arthropods is divided into cephalothorax and abdomen. Distinct carapace covers cephalothorax and most of exoskeleton is calcified. Each body segment possesses a pair of biramous appendages. Crustaceans can be distinguished from other groups of arthropods, such as insects by the possession of biramous (two-parted) limbs, and by the nauplius form of larvae.

There is almost 52,000 described crustacean species throughout the world, but the number of undescribed species may be 10-100 times higher (Martin andDavis, 2001). India has 2,934 crustacean species which is about 8.2% of total global crustacean fauna (Venkatraman and Krishnamoorthy, 1998). A total of 6 species belonging to 4 genera have been reported from Himachal Pradseh (Ramakrishna, 1995). Recently, Devroy (2008) has reported the presence of *Porcellionides pruinosus* as a new record from Pin Valley National Park.

Mollusca

Molluscs are soft-bodied animals, a majority of which are covered by a hard calcareous shell. The shell may consist of one, two or many pieces or sometimes may be internal and cartilaginous. Molluscs are structurally heterogeneous group since a slug is strictly different in structure from a fresh water mussel or from an octopus or a snail. The shell, by which majority of molluscs are known is absent in many forms. It includes the animals popularly known as snails, slugs, mussels, oysters, clams, cuttle fish, squids, octopuses etc. These animals occupy various habitats and are divided into marine, terrestrial and freshwater. Out of the seven classes of Mollusca, Gastropods and Bivalvia are abundant in fresh water, besides land, Gastropods also adapt more quickly to the environment.

Molluscs exhibit a great deal of adaptation to specialized ecological niches, making this group more vulnerable to modifications in the environment (Bouchet, 1996). Consequently, molluscs have suffered a severe decline in diversity and abundance due to human induced alteration of habitats, pollution, siltation, deforestation, poor agricultural practices, the destruction of riparian zones and invasion by exotic species (Biggins *et al.*, 1995). Land snails are an important component of the leaf-litter/soil biota and are responsible for nutrient recycling along with other soil invertebrates.

Mollusca are the second largest invertebrate group in abundance next only to Insecta. The estimated richness in this group varies from 80,000 to 135,000 species (Bruggen, 1995). Among these 31,000 to 100,000 species are marine, 14,000 to 35,000 terrestrial and about 5,000 are freshwater species (Seddon, 2000). Much of this diversity occurs in the tropical world. Still a large number of tropical molluscs remain to be described, partly because of under-exploration, and partly due to their minute sizes (Emberton, 1995 a; b, 1996). It has been estimated that in India there are some 5,000 species of these molluscans accounting to more than 7% of the total diversity of Mollusca (Rao, 1998).

Notable works on this group are *'Fauna of British India: Mollusca'* contributed by Blanford and Godwin-Austin (1908), Gude (1914, 1921) and Peterson (1915). A couple of compilation works on land snails have been published by the Zoological Survey of India. One publication has reviewed endemic land molluscs of India and another gives a brief description of selected land snails occurring in India (Ramakrishna and Mitra, 2002). Madhyastha *et al.* (2004) reviewed the status of land snails of the Western

Ghats and Mavinkuruve *et al.* (2004 a, b) made a provisional checklist of molluscs of Karnataka state. Apart from these studies, several papers appeared on the inventory of land snail diversity in several protected areas such as Rajiv Gandhi (Nagarahole) National Park(Ganesh *et al.*, 2002). Recently, effectiveness of chemo-attractive effects of amino acids against *Lymnaea acuminata* snails has been investigated (Tiwari, 2011).

Molluscs of Western Himalayas have been studied by a number of workers, some notable ones include Nevill (1878), Theobald (1878), Godwin-Austin (1899), Hora *et al.* (1955), Rajagopal and Subha Rao (1968; 1972), Davis *et al.* (1986) and Subba Rao and Mitra (1995). Works on molluscs of Himachal Pradesh include those of Hora (1928), Bhardwaj and Thakur (1973), Agrawal (1975 a ; b; 1976; 1977; 1979), Bhalla and Pawar (1977) and Rao and Mitra (2005).Patil (2008) reported three species namely *Lymnaea truncatula, Vallonia pulchella* and *Macrochlamys glauca* for the first time from Pin Valley National Park, Himachal Pradesh.

Odonata

Dragonflies and damselflies are amongst the most attractive creatures on the earth, the first to have conquered the aerial domain. The odonata are relatively large and often beautifully coloured insects that spend most of their time on wings. They are commonly known as damsel and dragonflies. The immature stages are aquatic while adults arefound usually near the water. Moreover, adults are large predacious insects while larvae are carnivorous and voracious feeders. Most of the species are generally beneficial. Nymphs develop rudimentary wing sheath in early stages and undergo incomplete metamorphosis.

Most species of odonates are highly specific to a habitat; some have adapted to urban areas and make use of man-made water bodies. Habitat specificity has an important bearing on the distribution and ecology of odonates. Odonata is represented by 6,000 species belonging to 630 genera in 28 families, clubbed under 3 suborders namely, Zygoptera, Anisozygoptera and Anisoptera from all over the world (Prasad, 1998).

In India, 499 species and subspecies under 139 genera in 17 families, 32 subfamilies and 7 superfamilies have been documented (Prasad and Varshney, 1995). In India, Fraser (1933, 1934, 1936), Prasad (1996) and Kulkarni and Prasad (2002) have left their impact on systematic of Odonata. Prasad and Kulkarni (2001) reported 71 species from Nilgiri Biosphere reserve. Further, Prasad and Kulkarni (2002) reported additional 34 species from Kerala. Shinde and Sathe (2006) recorded a total of 36 species of dragonflies from Koyna dam area (Western Ghats).

There is presence of distinct zoogeographical distributional patterns in the Himalayan fauna. The fauna of Western Himalaya is composed largely of Oriental elements, partly of Palaearctic and lesser of Mediterranean and Ethiopian elements (Mani, 1974). The dragonflies of western Himalayas have been catalogued by Kumar and Prasad (1981), and Kumar (1995). Similarly, in the recent past this group has been well studied in Himachal Pradesh by a number of field workers like Bhasin *et al.* (1953), Singh (1963), Kumar and Juneja (1976), Kumar (1978, 1982), Kumar and Prasad (1981) and Chandra (1983) who worked on Odonata fauna of different parts of Himachal Pradesh. Literature revealed the presence of 88 species under 52 genera belonging to Zygoptera (31 species) and Anisoptera (57 species) from Himachal Pradesh. Of these, 74 are Oriental, 11 Palaearctic, 02 Ethiopian and only 01 species is Circumtropical (Kumar, 2005).

Orthoptera

The Orthoptera is a group of large and easily recognizable insects like grasshoppers, locusts, crickets, mole crickets and grouse locusts. This name has been derived from Greek word *'Ortho'* and *'pteron'* meaning straight wing. The most significant feature of this group is its jumping habit with the help of large hind legs and sound production by its auditory organs. Orthoptera can also be excellent bio-indicators, as they can be recognised in the canopy at night without having resort to any trapping or landscape disturbance.

This is one of the largest insect orders with over 20,000 species known to the science throughout the world (Gillot, 2005). More than 1,750 species, about 10% of the total world species, have been recorded from India (Alfred and Ramakrishna, 2004). The studies on Orthoptera fauna of different parts of the world have been conducted by a number of workers like Dirsh (1956), Robertson (1967). Significant contributions to the Orthoptera were made in different parts of Asia by various workers who compiled the information on Tetrigidae from Mount Everest and Sri Lanka (Uvarov, 1926; 1928 a; b). Chopard (1969) reported it from Samoa, explored Sumatra and Sunda; Henry (1940) studied Orthopterans from Sri Lanka; Tinkham (1937) studied the Tetrigidea from Taiwan.

In India, the taxonomy of grasshoppers was initiated probably by Stal (1860, 1873). Thereafter, Walker (1870, 1871), Saussure (1884, 1888) and Bolivar (1902, 1909, 1914) studied the order Orthoptera of the Indian subcontinent. The most consolidated taxonomic work on Indian grasshoppers was made by Kirby (1914) in his notable work of 'Fauna of British India'. In the past, many workers like Walker (1869, 1871), Chopard (1969), Bhowmick (1985 a; b; 1986), Tandon and Shishodia (1969, 1995), Tandon (1976), Shishodia and Tandon (1993) have studied the Orthoptera of India. Similarly, many studies have been undertaken in India like those of Bolivar (1902), Uvarov (1929), Henry (1940), Tandon and Shishodia (1969, Tandon and Khera (1978) who have exclusively studied the grasshoppers and crickets of India.

Uvarov (1925) reported *Strauraderus biguttulus* (Linn.) from Kangra. Chopard (1969) in his monumental work 'Fauna of India' (Grylloidea) has reported the presence of some species of crickets like *Gryllotalpa fossor, Gryllus brunneri, Gryllus histricus* etc. from different parts of Himachal Pradesh. Bhowmick and Rui (1982) recorded the grasshopper fauna of Shivalik hills. Bhowmick and Halder (1983) reported 29 species of Acrididae from Himachal Pradesh. Bhowmick and Halder (1984) elucidated the distribution with some little known species of Acrididae from the western Himalaya (Himachal Pradesh). Bhowmick (1985 a) has outlined the distribution of Indian grasshoppers and recorded a total of 56 genera in family Acrididae. Bhowmik (1985 b) reported 70 species of Gryllid fauna belonging to 34 genera and 10 subfamilies from the Western Himalayas. Moreover, 21 species were recorded for the first time from the states of the Western Himalayas. Shishodia and Tandon (2000) compiled a list of 50 species of grasshoppers from Renuka Wetland,

Sirmour. Mehta *et al.* (2002 a) studied the Orthoptera of Kalatop-Khajjiar wildlife Sanctuary, Chamba. Shishodia *et al.* (2002) compiled the presence of 39 species belonging to 35 genera, 8 families and 4 superfamilies of grasshoppers from Pong Dam Wetland, Kangra. Thakur and Mattu (2006) reported 10 species of Orthoptera under 9 orders and one family (Acrididae) from Pin Valley National Park, Lahaul and Spiti, Himachal Pradesh. Saini and Mehta (2007) compiled the presence of 140 species under 2 suborders, 7 families and 88 genera from Himachal Pradesh. Recently, Shishodia and Gupta (2010) have corrected the above list and reported 165 species belonging to 105 genera and 16 families from Himachal Pradesh.

Diptera

These true flies include midges, mosquitoes, black flies, horse flies, house flies etc. They are distinguished from other insects by the presence of a single pair of patent, metathoracic flight wings (from the Greek *di* = two, and *ptera* = wings). One pair of functional wings is present on mesothorax while non functional halters are present on metathorax. Some authors draw a distinction in writing the common names of insects: true flies are written as two words, e.g., crane fly, robber fly, bee fly, moth fly, fruit fly. In contrast, common names of nondipteran insects that have "fly" in their names are written as one word, e.g., butterfly, stonefly, dragonfly, scorpionfly, sawfly, caddisfly, whitefly (Brown *et al.*, 2009). Dipterans are mostly minute to small soft bodied insects with highly mobile head, large compound eyes, variable antennae and suctorial type of mouth parts. They are fairly homogeneous in general appearance but some flies may be mistaken for some other kind of insects of no near relationship, e.g. Bombylidae, Syrphidae, etc., frequenting flowers as mimics of bees and wasps (Datta, 1998; Alfred and Ramakrishna, 2004).

There are more than one hundred thousand species belonging to some 130 families spread all over the world. Only six percent world dipteran diversity has been reported from India which is 6,093 species under 1,075 genera and 87 families (Datta, 1998). Dipterans have true worldwide distribution. Some species have been reported from Antartic also. It is one of the major insect orders both in terms of ecological and human (medical and economic) importance. The Diptera comprises an estimated 240,000 species of insects, although under half of these (about 120,000 species) have been described worldwide (Gillot, 2005).

All Diptera together in intimate association with other animals and variegated flora interact with each other in order to maintain the equilibrium of the nature. Flower visitors help plants perpetuate through pollination, larvae of some dipterans are phytophagous and of others annoy man and livestock. Dipterans show great diversity in habits and habitats. Their food comprises nectar, decaying organic matter and other insects while a few are blood sucker hence carrier of diseases like malaria, elephantiasis, sleeping sickness, kala azar and yellow fever etc.

Order Diptera is divided into three sub-orders- Nematocera possessing more than five segmented antenna, remaining two suborder possess less than five segmented antenna are as Brachycera (apical or subapical arista) and Cyclorrhapha (dorsal arista) (Parui and Mukharjee, 2000).

Nandi (2002 a) gave an account of the research done by various workers on Calliphoridae fauna of West Bengal. He reported 41 species of Calliphorid flies. The species of genera Calliphora, Chrysomyia, Lucilia and Hemipyrellia have diversified habitat and mostly are synanthropic, a few are forest habitants and attracted to garbages and dead bodies of vertebrates.

18 species of Syrphidae (out of total 25 species known so far) were reported from Jammu & Kashmir (Datta and Chakraborti, 1983). Their distribution was found wider as some were palaearctic and some oriental species. Singh and Maheshwari (1986) studied family Chironomidae as indicator of Lake Typology of Northwest Himalaya, India and concluded that Chandertal Lake is a zeta-oligotrophic lake. Eight new species of Chironomidae were reported from Chandertal lake in North West Himalaya. Palaearctic genus Corynoneura Winn.has also been reported for the first time from India along with three new species viz., C. carina*ta*, *C.chandertali* and *C. lahuli* (Singh and Maheshwari, 1987). They have also described five new species of genus *Himatendipes*.

New records of blow flies (Calliphoridae) from India viz. *Catapicephala ingens* (Walker), *Dexopollenia flova* (Aldrich), *Lucilia bazini* Seguyand *Isomyia tibialis* (Villeneuve) were recorded by Singh and Sidhu (2004 a, b). They compiled contribution of various workers viz. Crosskey (1965),James (1970; 1977), Kurahashi (1970; 1972),Pajni and Gera (1983) andSidhu and Singh (2002).Singh and Sidhu (2007) identified two new species of *Melinda* Robineau-Desvoidy (Diptera: Calliphoridae) from India.

So far, 364 species of Calliphoridae are known from Oriental region, including 101 species from India. More important contributions in this study of Indian blow flies (Calliphoridae: Diptera) have been made by Nandi (2004), Ronges (2009), Singh and Sidhu (2004 a; b). Biting midges of the genus *Palpomyia* meigen (Diptera: Ceratopogonidae) in India was studied by Gupta *et al.* (2008).

Dipteran fauna of Punjab and Himachal Shivalik hills has been explored by Parui *et al.* (2006) and 34 species belonging to 29 genera under 13 families have been reported. Studies on Dipterans of Himachal Pradesh are scanty. Brunetti (1917) made some contributions on the dipterans of Shimla, while Parui and Mukherjee (2000) have studied the diversity of dipeterans of Renuka Lake, Sirmour. 44 species of aquatic dance fly (Empididae) were identified from Himalayan stream system of Kullu valley of Himachal Pradesh. Out of these 44 species, 33 species were identified to species level and 17 species were new to literature (Wanger *et al.*, 2004).

Hymenoptera

This order is probably the most beneficial in the entire insect class. It contains many insects that are of value as parasites or predators of various insect pests, and contains the most important insects (the bees) that are involved in pollination of plants. However, a relatively small number of them act as pest of crop and forests.

Hymenopteran can be studied in three categories depending upon their habits. The primitive ones are phytophagus hence lays eggs inside plant tissue from which their larvae get nourishment (Cynipidae). The second category includes social insects like ants, wasps and bees which establish their own colonies leading a complex, social and co-operative life. The last category is known as entomophagous insects which include parasites and predators (Jonathan, 1998).

Scientists have estimated near about 200,000 species throughout the world but only 120,000 species have yet been described. Out of these about 20,000 species belong to superfamily Apodea while parasitic family Ichneumonidae contains about 60,000 species. Gillot (2005) reported 130,000 hymeopteran species in world. India holds about 8.3% of total world's hymenoptera with about 10,000 species (Jonathan, 1998).

Among all orders of insects this order stands out for its typical way of living on the ground, utilising the environment fully and at the same time controlling other insects. Insects of this order are distributed throughout the world showing their capacity to use various environmental conditions. About 18,000 species have been reported from North America, 15,000 species from Australia and 6500 species from the United Kingdom. They have developed complex reproductive behaviour pattern, particularly related to provision of food to progeny, hence evolution of sociality in its many groups (Gillot, 2005). Metamorphosis is complete and in most of the families larvae are grub like or maggot like. Hymenopterans are distributed throughout India, and data clearly shows that the affinities of Indian taxa are mostly with the fauna of Oriental Region. The Himalayan region harbours about 60 percent of species of Hymenopterans which shows affinities with the taxa of Palaearctic region. The South Indian taxa resembles with Srilankan taxa while those of Andaman and Nicobar Islands are close to Indonesian taxa. About 20 percent taxa present in hilly region of Eastern and Western Ghat while remaining 20 percent in Central and North India (Jonathan, 1998). Out of about 1200 species or subspecies of family Ichneumonidae about 250 species are discovered during the last 15 years. About 25 percent of Hymenopteran species are still to be described in India. Hymenopterans are divided into two sub-orders: Symphyta have 4 super families and Apocrita with 11 super families.

Bingham (1897; 1903) for the first time gave full account of Wasps and Bees in two volumes of his 'Fauna of British India, Hymenoptera'. Morely (1913) published an account of family Ichneumonidae in Volume III of 'Fauna of British India, Hymenoptera'. Betrem (1928) has worked on family Scoliidae. Krombein (1982) published monographs on the families Scoliidae and Tiphiidae from Sri Lanka.

Gupta (1994) reported 14 species of Tiphiiidae and Scoliidae belonging to three subfamilies namely Myzininae Typhiinae and Scoliinae under six genera from Rajaji National Park, India in which six species namely *Mesa petiolata, M. clariopensis, Tiphia coimbatorea, T. tequlita, T. descreascens* and *T. palmi* were new record for Uttar Pradesh. Gupta (1995) studied and published Hymenopteran fauna of western Himalayas from eight districts of Uttar Pradesh (Kumaoan and Garwal region). He described the distribution of about 200 species belonging to 12 families. From Nanda Devi Biosphere Reserve, Gupta (1997) gave detailed systematic account of 24 species of Hymenoptera. The genus *Mellinus fabricius* (Sphecidae) was recorded for first time from Oriental region. The total number of species from Biosphere Reserve rose to 28. Chavhan and

Pawar (2011) reported 34 species of ants belonging to 20 genera in forest, grassland and human habitats located around Amravati city.

An account of 25 species belonging to six families was recorded from Gobind Pashu Vihar, Uttarakhand. Out of which *Odontomachus punctulatus* Forel (Formicidae) and *Odynerus sikkimiensis* Bingham (Eumenidae) were new distributional records from Western Himalaya (Gupta, 2004). Seven species of Bumble bees (Bombidae) were recorded along with their host plants from Lahaul and Spiti region of Himachal Pradesh by Saini and Ghattor (2007). Hymenoptera fauna of Pin Valley National Park in Lahaul and Spiti district of Himachal Pradesh has been studied in detail by Gupta (2008) describing seven species belonging to families Vespidae (4 spp.) and Sphecidae (3 spp.). Out of seven species, three species of Vespidae (*Potistes associus, Dolichovespula asiatica* and *Vespula germanica*) and three species of Sphecidae (*Bembix latitarsus* and two species belonging to genus *Podalonia*) were new records from Himachal Pradesh. *Podalonia* species are common in Lahaul and Spiti district and known as ectoparasitoids of lepidopteran larvae. Moreover, two species of Torymidae under the genera *Ecdamua* Walker (subfamily: Toryminae) and *Megastigmus* Dalman (subfamily: Megastigminae) have been identified as new to science from Ladakh (**Sureshan, 2009**).

Lepidoptera

Butterflies and moths are grouped in this order. Order Lepidoptera is divided into two sub-orders, the Rhopalocera and the Heterocera. Butterflies are placed in the Rhopalocera while the moths are placed in the sub-order Heterocera. They are most readily recognised by the scales on wings. The members are of great value both for conservation and for environmental planning in local scale. The members of this order undergo complete metamorphosis, and their larvae, usually called caterpillars is a familiar sight (Borror *et al.*, 1981).

Hampson (1918) estimated as many as 89 families and subfamilies of Lepidoptera, while Hamlyn (1969) reported about 140,000 species comprising 13,000 butterflies and 1,27,000 moths from all over the world. A recent estimate shows the occurrence of about 1,42,500 species of Lepidoptera around the globe, but estimates within Lepidoptera from the Indian sub-continent reveal that the group comprises over 15,000 species and many more subspecies distributed over 84 families and 18 superfamilies (Alfred *et al.*, 1998). There are about 1500 species of butterflies in India (Gay *et al.*, 1992).

Rhopalocera (Butterflies): Out of 84 families and 18 superfamilies of the order found in the Indian subregion, the butterflies belong to 5 major families placed under 2 superfamilies, these constitute about 10% of the total faunal species. A total of 123 species of butterflies are considered endangered in India (Mondal, 1998).

Heterocera (Moths): Moths are easily distinguished from butterflies due to the presence of thread like antennae, wings folded in a roof-like manner over the abdomen and nocturnal habits.

Lepidopterans are distributed in a wide variety of habitats. It has been revealed that the niches and habitats of Lepidoptera sprawl from the mountains to the mangroves and also the major insular belts like the Andaman, Nicobar and the Lakshadweep islands on the basis of eco-faunistic surveys made at different localities of India by various workers including those of the Zoological Survey of India. Maximum degree of faunistic stagnation is found in the East Himalaya where as in southern and insular parts of India, the number of lepidopteran species is proportionately less. Many more species are, however, yet to be explored from the remote corners of plains, arid and wastelands and forest covers in different areas from East to West and North to South India (Alfred *et al.*, 1998).

Out of 84 families and 18 super families of Lepidoptera available in Indian sub-region, the butterflies belong to five major families viz., Papilionidae, Pieridae, Nymphalidae, Lycaenidae and Hesperiidae under two superfamilies viz., Papilionoidea (including the first four families) and Hesperiioidea (including the last family) and constitute about 10% of the total faunal species. The largest representative families of butterflies and moths from India are Nymphalidae (450 species) and Noctuidae (1,500 species), respectively. A few families of moths like Castniidae and Neopseustidae are very poorly known from the country (Varshney, 1993).

The lepidopteran fauna, particularly the Rhopalocera or the butterflies, attracted attention of both the naturalists and zoologists for their aesthetic as well as scientific values. During the eighteenth century, a large number of butterfly fauna from the Indian region, including those from theWestern Himalaya, was named and described by Linnaeus and Fabricius (Arora *et al.*, 2005).

The other major contributions on diversity and distribution of butterflies were provided by Marshall and de Niceville (1883; 1886; 1890) and de Niceville (1886) from India, Sri Lanka and Doherty (1886) from Kumaon and Mackinnon and de Niceville (1897, 1898) from Mussorie, India. Valuable contribution of Bingham (1905, 1907) in the form of two volumes "The Fauna of British India, Butterflies' and Swinhoe's (1909-1913) "Lepidoptera Indica" Vols. VII-X, are worth mentioning. Evans contributed a series of papers on the "Identification of Indian Butterflies". These were later published (1932) in the form of a book entitled "Butterflies of Plains of India". While Evans (1949) published a catalogue of Hesperiidae from Europe, Asia and Australia; Cantile (1962) studied the Arthopola group of Lycaenidae.

Wynter-Blyth (1957) published a book entitled 'Butterflies of the Indian Region'. Besides these, the studies conducted by Hemming (1967) of the Indian butterfly fauna are a valuable contribution. Many scientists conducted studies on Lepidoptera (Rhopalocera) in different parts of India like Larsen (1987) in Nilgiri Mountains; Nandi (1987) studied the Lepidoptera fauna of Orissa; Gupta and Shukla (1987; 1988) in Bastar; Gupta and Thakur (1990) in Gujarat are some of notable ones. Ghosh *et al.* (1991) reviewed the distribution of some butterflies in different regions of India. Mandal and Maulik (1991; 1998) in Orissa and Meghalaya; Nandi (1993) from Orissa; Mathew and Ramathulla (1993) from Silent Valley National Park, Kerala; Arora (1994; 1995) from Rajaji National Park and Western Himalaya. Ghosh and Chaudhary (1997 a; b) studied the taxonomic composition of 39 species of Satyridae, 32 species of Hesperiidae and 25 species of Pieridae in West Bengal. Gupta (1997 a; b; c) studied the taxonomic account of butterfly in Delhi and West Bengal; Mandal *et al.* (1997 a) from Delhi; Khatri (1997) from Nicobar; Ghosh and Chaudhary (1998 a, b) and Mandal and Maulik (1998) from Meghalaya; Arora (2000) from Renuka wetland; Mandal *et al.* (2000) from Tripura; Andheria (2001) from Sanjay Gandhi National Park (Mumbai); Haribal (2001) studied over population of *Danaus genutia* in Mumbai; Pai (2002) studied the butterfly distribution pattern in Goa; Radhakrishnan and Sharma (2002) in Eravikulum National Park, Kerala.

Mehta *et al.* (2002 b) studied butterflies of Pong Dam wetland in District Kangra (H.P.) and reported 50 species belonging to 37 genera under seven families. Moreover, a preliminary study on butterflies of Himachal Pradesh as flower visitors and pollinators was carried by Mehta *et al.* (2003). A total of 67 species of Butterflies belonging to nine families were reported from Sukhna lake and catchment area (Chandigarh) by Thakur *et al.* (2006 a).

Distributional records of Rhopalocera from Pin Valley National Park were studied and 14 species belonging to 11 genera and four families were reported (Thakur and Mehta, 2006). Role of butterflies in pollination in Shivalik hills has recently been studied by Thakur and Mattu (2010). Walia (2005) has reported 184 species of moths (family: Geometridae) from Himachal Pradesh.

Coleoptera

The order Coleoptera is the largest order of insects and contains about 40 percent of the known species of class hexapoda. Insects of this order are commonly known as beetles. One of the most distinctive features of coleoptera is the structure of wing. Their name is derived from two Greek words meaning sheathed wings. The beetles undergo complete metamorphosis, larvae has different forms in different families.

The diversity of beetles is very wide-ranging. They are found in almost all types of habitats, but are not known to occur in the sea or in the Polar Regions. They interact with their ecosystem in several ways. They often feed on fungi, break down animal and plant debris, and eat other invertebrates. Some species are prey of various vertebrates including birds and mammals. Certain species are agricultural pests and other plant-sucking insects that damage crops. There are over 1,000 known species of beetles to be either parasitic, predatory or commensals in the nests of ant (Neumann and Elzen, 2004; Peck, 2006).

Coleoptera is most diverse order of class Insecta. One out of every four described animal species is beetle. There are 350,000 species belonging to 177 families under four suborder estimated by scholars throughout the world. But still it is doubtful whether this number represents one-tenth of those existing today. The largest taxonomic family, the Curculionidae (the weevils or snout beetles), also belongs to this order. Some 70%-95% of all beetle species, depending on the estimate, remain undescribed. From India about 15,500 species belonging to 104 families under three sub orders have been reported. If good exploration of habitats and inventorisation of collections are made then the number of species would at least double to the reported figure (Sengupta and Pal, 1998).

Recently, a new jewel beetle genus, with one species (*Cretofrontolina kzyldzharica*) from the Upper Cretaceous of Kazakhstan was described based on a body; and three new species of the formal genus *Metabuprestium* were described based on isolated elytra from Upper Cretaceous of Russia and Lower Cretaceous of Mongolia (Alexeev, 2009). Indian Coleoptera have been dealt and described by various workers like Westwood (1832) and Hope (1837) who recorded 5000 species from different parts of India. Lefroy and Howlet (1909) who studied Indian insect pests and Stebbing (1914) explored ecology and control of Indian forest insects.

Many workers contributed for the 'Fauna of British India' series, like Gahan (1906) on Cerambycidae; Jacoby (1908) on Chrysomedlidae; Arrow (1910; 1931) on Scarabaeidae; Fowler (1912) on Cicindellidae, Rhysodidae and Paaussidae and Marshall (1916) on Curculionidae family. Basu (1986) described the fauna of Chrysomelidae family from the Silent Valley, Kerala. Ghosh *et al* (1991) reported 25 species of coccinellids which prey upon 52 species of Aphids infesting 45 different host plants. Basu and Halder (1987) described the state fauna of Chrysomelidae of Orissa. Singh (2001) reported 11 species of Coleoptera from Sukhna lake and Catchment area. Saraswat (2002) and Mehta (2003) reported seven and nine species respectively belonging to different families including Coccinellidae from Shimla hills. Poorni (2003) described new species of Coccinellidae family from Himachal Pradesh. Communal roosts of three species of tiger beetles (*Colomera angulata*, *Colomera plumigera* and *Colomera chloris*) were studied by Bhargav and Uniyal (2008) in the Shivalik Hills of Himachal Pradesh. Nine species of pleurostic Scrarbaeidae were identified by Chandra and Uniyal (2007) from the Great Himalayan National Park, Himachal Pradesh.

Pisces

This group of animals is formed of aquatic cold-blooded vertebrates possessing gills for respiration throughout life and limbs, if any, are in the shape of fins and are primarily depend on water as a medium to live in. The living fishes are divided into 4 classes, out of which 2 classes, viz., Chondrichthyes (those with cartilaginous skeleton) and Osteichthyes (those with bony skeleton) are represented in India. This group comprises about half the total number of vertebrates. Nelson (2006) estimated 27,977 valid species of fishes under 62 orders, 515 families and 4,494 genera in the world and the eventual number of extant fish species was projected to be close to 32,500. About 11,952 species or 42.72%, normally live in freshwater lakes and rivers that cover only 1% of the earth's surface and account for a little less than 0.01% of its water. The secondary freshwater species number 12,457 and the remaining 3,568 species are exclusively marine. Jayaram (1999) listed 852 freshwater species of fishes under 272 genera, 71 families and 16 orders, including both primary and secondary freshwater fishes from India, Bangladesh, Myanmar, Nepal, Pakistan and Sri Lanka.

The Indian species represent about 8.9% of the known fish species of the world. Talwar and Jhingram (1991) estimated 2,546 species of fish belonging to 969 genera, 254 families and 40 orders in the Indian region. Chondrichthyes are represented by 131 species belonging to 67 genera, 28 families and 10 orders, and Osteichthyes by 2,415 species in 902 genera, 226 families and 30 orders (Barman, 1998). The checklist of Menon (1999) lists 446 primary freshwater species under 33 families and 11 orders from the Indian region alone. Of the primary freshwater species 68% are constituted by the Cyprinoids, 18% by Siluroids and 14% by other groups.Important contribution in study of fishes in India has been made byJayaram (1999, 2006), Menon (1999), Ponniah and Gopalakrishnan (2000), Ponniah and Sarkar (2000), Day (1889 a; b)and Vishwanath *et al.* (2007).

Studies on fish fauna of different Indian states have been conducted by some workers in the recent past like Vats and Gupta (2011) who reported 64 species belonging to 35 genera and 16 families from four districts of northern Haryana.Sharma andMehta (2009) reported 19 species of fish fauna extended over 10 genera and three families from rivers of Ladakh.

Studies on the fish fauna of Himachal Pradesh have been carried out by various workers,the most important contributors being McClelland (1839, 1842), Stenidachner (1867), Hora (1927), Menon (1951, 1954, 1962, 1974, 1987, 1999), Bhatanagar (1973), Tilak and Husain (1977), Sharma and Tandon (1990), Mehta (2000 a), Johal *et al.* (2002), and Dhanze and Dhanze (2004). Mehta and Uniyal (2005) reported 104 species of fish belonging to 48 genera, 14 families and 8 orders from all 12 districts of Himachal Pradesh. They have recorded maximum number of species, i.e., 57 from Sirmour district, followed by 55 from Kangra and 50 from Bilaspur. Minimum number of species (2 species) was from Lahaul and Spiti district. Moreover, Mehta and Sharma (2008) have reported three species of fish from Pin Valley National Park, Lahaul and Spiti of Himachal Pradesh. Study on the hydrological conditions of river Beas and its fish fauna in Kullu Valley has been conducted by Kumar (2010) and revealed the presence of 6 species of fish belonging to 3 orders and 3 families.

Amphibia

Amphibians are poikilothermic (cold blooded) vertebrates with smooth skin leading a bimodal life, i.e., life in water as well as land (Amphi- meaning "on both sides" and -bios meaning "life"). This phylum includes animals such as toads, salamanders, caecilians and frogs. They may or may not possess limbs and tails. Much of gas exchange in them occurs through their skin whose secretion protects it from desiccation (Chanda, 1998). The three modern orders of amphibians are Anura (tailless and limbless animals like toads and frogs), Caudata (tailless animals e.g. salamanders and newts), and Gymnophiona (caecilians, limbless amphibians that resemble snakes).

Amphibians are least harmful in nature and are found throughout world from sea level to an altitude of about 3,500 m. They do not cause any depredation to agriculture crops, fruits and vegetables. On the contrary, their food mainly consists of small insects and their larvae, algae, snails etc. which are pest of crops and vectors of some diseases. Amphibians are ecological indicators and in recent decades there has been a dramatic decrease in their number. Many species are now threatened or have become extinct.

They are the least amongst the vertebrates and comprise nearly 6.6% of the total vertebrate life on earth (Lagler *et al.*, 1962). Total number of species in the world has been estimated around 3,140 and in India 214 species are known, while in Himachal Pradesh only 17 species belonging to 4 families have been recorded. This is 7.8% of the total Indian species (Mehta, 2005). The information on the amphibian fauna of different parts of the state is available in the works of Annandale (1907), Boulenger

(1920), Kriplani (1952), Dubois (1975), Tilak and Mehta (1983) and Mehta (2000 b, 2005). Mehta (2009) has reported three species of amphibian (*Pseudopidalea latastii, Bufo viridis* and *Scutiger occidentalis*) from Indian Cold desert Ladakh.

India has the third largest amphibian population in Asia. The amphibian fauna of India comprises of 214 species of which 167 (66.3%) are endemic to the country. In spite of its broad variety of species, India holds second place on the list of countries having the most number of threatened amphibian species in Asia, with 67 (25%) of its species facing possible extinction. Some of the important contribution to amphibian research in India has beenmade by Smith (1935), Gruber, (1981), Dutta (1997), Chanda, (2002) and Dinesh *et al.*, (2009).

Reptilia

Reptiles are cold-blooded vertebrates which breathe by the lungs throughout their life and their body remains covered with scales. They are tetra-pods, either having four limbs or being descended from four-limbed ancestors. Unlike amphibians, reptiles do not have an aquatic larval stage. As a rule, reptiles are oviparous (egg-laying), although certain species of squamates retain the eggs until hatching and a few are viviparous.

Reptiles were the dominant group of vertebrates during the Mesozoic period and reached their maximum and most diversified development at that time. Most of the orders of reptiles were established by the end of Triassic and also some became extinct at that time. Reptiles originated around 320-310 million years ago during the Carboniferous period. Of the 19 orders of reptiles, only 4 survive today and they are typically recognized. Of these, Crocodilia includes crocodiles, gavials, caimans, and alligators having 23 species. The second order is Sphenodontia having two species found in New Zealand. The third order is Squamata which includes lizards, snakes and worm lizards having approximately 9,150 species. The last order is Testudines which includes turtle, terrapins and tortoises having over 300 species.

Reptiles are diverse in south Asia with approximate 632 species belonging to 185 genera and 25 families. India harbours 456 species of reptiles belonging to 25 families and 4 orders including 3 species of Crocodilia, 31 of Testudines, 178 of lizards and 244 species of serpents.

Das *et al.* (2009) have studied reptilian diversity of Barail Wildlife Sanctuary and adjacent regions of Assam in northeast India. There search by Das and Saikia (2007) in Barpete and its surroundings areas of Assam revealed the presence of four Trinonychidae species, i.e.,*Aspideretes gangeticus, Aspideretes hurum, Chitrs indica* and *Lissemys punctata*. However, a total of 35 species of reptiles belonging to three orders, 11 families and 28 genera were recorded from Pura Wildlife Sanctuary of Gujarat (Vyas, 2007). Other recent works on reptiles in different parts of India are those of Sharma (2005), Khaire (2006), Gomathi and Singh (2007), and Nameer *et al.* (2007). Recently, Venugopal (2010) has revealed the presence of 199 species of lizards within the boundaries of India.

Reptiles of Himachal Pradesh have been studied well during the fBritish period and a number of workers have explored the reptilian fauna of different areas of the state. In the 'Fauna of British India' Smith (1935; 1943) reported seven species of reptiles from various parts of Himachal Pradesh. A few workers like Acharjee and Kripalini (1951), and Waltner (1974) studied geographical and altitudinal distribution of reptile fauna along the Himalayas including Himachal Pradesh. Mehta (2000) reported 14 reptile species from Renuka wetland area. Saikiaand Sharma (2009) reported 17 species belonging to 10 families from Simbalbara Wildlife Sanctuary. Out of these, two species of lizards namely *Ophisops jerdoni* and *Eurylepis taeniolatus* and one species of turtle *Melanochelys trijuga indopeninsularis* recorded from the sanctuary were the first records for the State of Himachal Pradesh (Saikia *et al.*, 2010 b). A list of 55 species of reptiles belonging to 40 genera and 14 families has been prepared from Himachal Pradesh by Saikia *et al.* (2007). Some other useful works on the reptilian fauna of Himachal Pradesh are those of Saikia *et al.* (2008) who studied the Herpetofauna of Pin Valley National Park, Saikia and Mehta (2009) elucidated the reptilian diversity of Pong Dam Wetland, Saikia *et al.* (2010 c) who for the first time recorded the large worm snake (*Typhlops diardii*) in Himachal Pradesh.

Aves

Birds are one of the most fascinating creatures of nature which are cosmopolitan in distribution and are of great importance to mankind. They are characterised by the presence of feathers which serve many purposes like insulating the body and contributing to the flying apparatus of wing and tail. Birds are good indicators of biodiversity, and are a measure of the sustainability of human utilization of the natural environment.

The World List of Living Birds computes about 9,026 species under 1800 genera, 182 families and 30 orders (Saha, 1998). At present, India has a total of 74 restricted range species, of which 39 are confined to the geographical boundaries of the country (Stattersfield *et al.*, 1998). Besides, 79 Indian bird species are globally threatened with extinction. Further, more than 52 are classified as Near Threatened (Bird Life International, 2001). A large proportion of the rest of the bird species in India are rapidly declining and in urgent need of conservation (Jhunjhunwala *et al.*, 2001). Around 1,300 species of breeding, staging and wintering birds, spread over 88 families and 22 orders, occupying a wide array of natural, semi-natural and urban habitats are known from India (Manakadan and Pittie, 2001). It corresponds to as many as 13% of the world's birds.

Avifauna of various parts of the world has been explored by a number of workers like semi-arid area of Kenya has been explored by Pomeroy and Tengecho (1986), birds of Gola forest, Sierra Leone by Allport *et al.* (1988), and correlation of bird diversity with topography, precipitation, and an interaction between topography and latitude in South America by Rahbek and Graves (2001). In a study from coastal forest of Ubatuba in southeast Brazil, 114 bird species were recorded (Dario and Vincenzo, 2011). Similarly, abundance–occupancy patterns and sensitivity to forest fragmentation for avifauna were conducted by Anjos *et al.* (2011) in three distinct forest ecosystems found in Brazilian Atlantic forest. In India as well, some extensive studies have been undertaken. Daniel (1989) inferred various conservation measures, based upon the studies on birds of Uttara Kannada district of Karnataka. Moreover,

37 species of birds were recorded from Morana lake of Sangli district of Madhya Pradesh (Abdar *et al.*, 2006). Similarly, 23 species of birds were recorded from mangroves of Akshi creek, Alibaug (Yeragi and Yeragi, 2006). A Study was conducted to assess the winter migrant birds, especially water birds population in Nelapattu Bird sanctuary in Nellore district of Andra Pradesh. A total of 37 species were reported belonging to 14 families (Guptha *et al.*, 2011).

Studies on the avian diversity of Ropar Wetland, Punjab by Mehta *et al.* (2002 b) revealed the presence of 206 species of residential, winter and summer visitors belonging to 152 genera, 50 families and 17 orders. Similar studies on different aspects of birdlife have been undertaken in different parts of Punjab by some workers like Reeves (1981) who studied the birds of Sukhna Lake, Chandigarh; Khera *et al.* (1984) explored avian fauna of Chandigarh; Kalsi and Sodhi (1986) studied some birds of Chandigarh and surrounding areas; Narang (1986) recorded birds of Sukhna Lake, Chandigarh; Dhindsa, *et al.* (1991) added some new birds to the checklist of birds of Punjab; Kazmierczak and Singh (1998) portrayed the picture of birdlife of North India; Gopi Sundar (2000) discussed the distribution, demography and conservation status of Sarus Crane in North India; Singh (2001) illustrated the birdlife of Harike Wetland, Punjab. Recently, Thakur and Paliwal (2012) have reported the presence of 239 species of birds belonging to 156 genera, 54 families and 17 orders from Chandigarh.

Bird diversity studies in the Himalaya have always fascinated a number of ornithologists. Ladakh area of Jammu and Kashmir has always been an area of special attraction, as a few species of water birds in Indian subcontinent breed in the lakes and marshes of this area in winter. Secondly, it was the first part of the Tibetan plateau to be opened to the outsiders. Several bird collecting expeditions explored Ladakh area especially those by Stoliczka (1868), Ludlow (1920), Osmaston (1926). Holmes (1986) documented a total of 128 species of birds from Suru valley of this region.

The earliest record on the study of birds of Himachal Pradesh pertains to Theobald (1862), Tytler (1868), Anderson (1889), Dodsworth (1910 a; b; 1911 a; b; c 1912 a; b; c; d; e; f; g; 1913 a; b; 1914). Studies by Whistler (1926 a; b) on birds of Kangra and Kullu districts are excellent accounts of representative areas. Frome (1946) studied avifauna of Mahasu Narkanda Baghi area. Ali (1949) has listed about 225 birds from Western Himalayan region. Wynter-Blyth (1951, 1952) portrayed the vision of a naturalist in the northwest Himalaya and described the birds of Himachal Pradesh especially Kullu, Manali, Parbati Valley and Spiti areas. Lavkumar (1962) gave a very interesting picture of birdwatching in the Northwest Himalaya. It was Ganguli (1967) who studied the birds of Shimla during autumn while Mistry (1967) studied the diversity and distribution of birds between Shimla and Kullu areas.

The bird diversity of Mandi district of the State has been studied by Mahabal and Mukherjee (1991). Further, Suyal (1992) made a valuable contribution to the data on birds of Himachal Pradesh by studying the birds of Rampur Bushar and Sarahan area of Shimla district. Similarly, Mahabal and Sharma (1992) reported the presence of 133 species of birds belonging to 46 families and subfamilies alongwith their altitude wise abundance in different zones of Kangra district of Himachal Pradesh. Narang and Singh (1995) studied the birds of Nauni campus of University of Horticulture and Forestry, Solan. Moreover, bird survey by Mahabal (1996) in Shivalik Himalaya covering the district of Una, Hamirpur and Bilaspur revealed the presence of 136 species belonging to 41 families alongwith their status and distribution range. Sharma and Mahabal (1997) conducted studies on seasonal changes in bird species in two different altitudinal locations (seasonal altitudinal migration) of Solan district of Himachal Pradesh. Similar studies on diversity, status and seasonal altitudinal movement in birds of Renuka wetland of the state have been conducted by Mahabal (2000) and Bhardwaj (2003).

Besten (2004) studied and compiled information on the birdlife of Kangra district of Himachal Pradesh and recorded 555 species from this area which included the only Ramsar Site of the State i.e. Pong Wetland. Further, Mahabal (2005) has enlisted a total of 447 species of birds from different ecogeographic zones of Himachal Pradesh. Gaston and Singh (1980) studied the status of Cheer Pheasant Catreus wallichi in the Chail Wildlife Sanctuary of Himachal Pradesh. It was the work of Gaston *et al.* (1981) which firstly threw sufficient light on the distribution and status of pheasants in Himachal Pradesh. Garson (1983) updated the information on Cheer Pheasants in Himachal Pradesh.

Mukherjee and Chandra (1984) enlisted the birds of Silli Wildlife Sanctuary, Solan. Gaston and Pandey (1987) sighted Red-necked Grebe (*Podiceps grisegena*) on the Pong Dam Lake. Pandey (1989 a) made some observations on the birds of Pin Valley National Park, Spiti. Further, Pandey (1989 b) recorded more than 220 species of resident, winter and summer visitor birds from the Pong Dam Wetland, Kangra. Moreover, Singh *et al.* (1990) enlisted about 358 birds from 2 National Parks and 29 Wildlife Sanctuaries of Himachal Pradesh. Gaston *et al.* (1993) reported the presence of 183 species of birds, including 132 passerines from the Great Himalayan National Park of Himachal Pradesh. Sangha (2005) added some new and significant bird species from the Great Himalayan National Park of District Kullu, Himachal Pradesh. Thakur *et al.* (2006 b) reported the presence of 92 species of birds from Tara Devi, Shimla. Recently, Thakur and Mattu (2012) carried out studies on diversity and distribution of avifauna of Himachal Pradesh which revealed the presence of 322 species belonging to 190 genera under 60 families and 17 orders. In different biogeographical zones of state, family Muscicapidae represented maximum number of species (67 spp.), followed by Accipitridae (22 spp.), Anatidae (17), Scolopacidae (13), Corvidae (12), Fringillidae (11), Phasianidae, Columbidae and Picidae (10 spp. each), Motacillidae (9) Ardidae (8), Charadriidae, Laridae and Hirundinidae (7 spp. each) 19 families including Upupidae were represented by single species each. Sharma *et al.* (2009) revealed the presence of 210 species distributed over 15 orders and 46 families in Simbalbara Wildlife Sanctuary.

Studies on avifauna of Greater and Trans Himalayan region of Himachal Pradesh have been undertaken by some research workers like Wynter-Blyth (1948, 1952) who studied the birds of Sangla valley of Kinnaur District and Spiti Valley of Lahaul and Spiti district. Mahajan and Mukherjee (1974) made some observations on the birds of Lahaul and Spiti.

Narang (1989) recorded the birds of Sangla valley (Kinnaur), while Pandey (1989 a) studied the birds of the Pin Valley National Park, Spiti. Besides, there are some interesting observations on different aspects of birdlife made by different workers viz., Khacher (1986) reported migration of Tufted Duck on Rohtang Pass, while Rana (1997) recorded the presence of Pallas' Fishing Eagle from Spiti valley; Manjrekar and Mehta (1999) reported Pond Heron in Pin Valley National Park, Spiti; Singh (2003) explored the birds of Tabo, Spiti; Santharam (2005) recorded the birds of a trek in the Chansal Pass. Recently, studies on birds in certain areas of Mandi and Solan districts have been reported by Thakur *et al.* (2010 a, b).

Some avifaunal studies have been undertaken in different parts of Chamba district of Himachal Pradesh like Tak (1987) who reported a rare sighting of threatened Western Tragopan (*Tragopan melanocephalus*) in Chamba district of the state, Mahabal (1992 a; b) who explored the avifauna of Chamba district and recorded the extension in distribution range of a few species of birds, and Pandey *et al.* (2004) analysed the birdlife of Kalatop-Khajjiar Sanctuary and based on these records, the sanctuary was declared as an Important Bird Area.

Mammalia

Mammals are warm blooded, viviparous animals, and are distinguished from birds, as well as from other vertebrate animals, by possession of mammary glands, secreting a nutritious fluid called milk for the nourishment of their young ones. They are also distinguished by covering of hair, entire or partial, at least during some period of their life cycle. Mammals are distinct vertebrates with adaptive plasticity as they widely exploit the resources of earth from one pole to the other and mountain top to deep ocean.

The global mammalian fauna is represented by 4,629 species under 1135 genera, 136 families and 26 orders (Wilson and Reeder, 1993). Small mammals form the highest proportion of mammals all over the world (Takele *et al.*, 2011). Among them, the order Rodentia has more species. Rodents account for more than 40% of the mammalian species in the world with 21 living families, 443 genera and more than 2,000 species.

390 species of mammals belonging to 180 genera, 42 families and 13 orders are found in India (Agrawal, 1998). Out of 180 genera, 61 are monotypic and 105 are represented in our country by a single species. Of the 390 species, 175 are threatened with extinction to various levels, and on that basis 75 have been listed in Schedule I, 73 in Schedule II, 8 in Schedule III and 19 in Schedule IV of the WildLife (Protection) Act 1972 (Agrawal, 1998). Chakraborty *et al.* (2005) has reported 107 mammalian species belonging to 77 genera, 25 families and 9 orders from Himachal Pradesh. Despite smaller area of only 1.7% of total geographical area of India, the State harbours 27% of mammalian species of India. Out of 107 species of mammals found in the state, 21 have been included in Schedule I of the Indian Wildlife (Protection) Act, 1972. Recently Sharma and Saikia (2009) has updated the information on mammalian fauna of Himachal Pradesh and reported the presence of 111 species.

Mammalian fauna of Himachal Pradesh is an admixture of Palaearctic and oriental elements since the state lies in transition zone of two biogeographical realms.

Brown Bear, Lynx, Alpine Weasel, Mountain Noctule etc. are some of Palaearctic representatives of State which probably reached from Hindukush Mountain and Russian Uzbekistan (Roberts, 1977). Some of the representatives of Oriental fauna of the State include Leopard Cat, Yellow Throat Marten, Himalayan Palm Civet, Indian Pangolin, Grey Goral, Barking Deer, Bandicoot Rat, Bush Rat, Flying Fox, False Vampire, Fulvous Leaf-nosed bat, Musk Shrew etc.

Mammals exploit almost all types of habitat, from snow covered Himalayas to plains, from thick rain forests to arid regions and terrestrial to aquatic realm. The mammalian fauna is rich in warmer part of Palaearctic region. There are about 37 races of Mountain sheep alive (23 Asiatic and 14 American races) today. Many mammalian species found are exclusively in mountains of Palaearctic region like Siberian and Alpine Ibex etc. These were originally plain living animals but took refuge in the mountains after Pleistocene era (Tiwari, 2007).

Sharma *et al.* (2006) has studied feeding behaviour of Black buck (*Antelope cervicapra*) in Gangetic plains of Bihar. Summer feeding habits of the Indian Giant Flying Squirrel *Petaurista philippensis* was studied in Sitamata Wildlife Sanctuary by Bhatnagar *et al.* (2010). Feeding, ecology and distribution of Himalayan Serow (*Capricornis thar*) in Annapurna Conservation Area, Nepal was studied by Giri *et al.* (2011).

The first report pertaining to the Chiroptera of Himachal Pradesh was that of Dobson (1873) who described *Vespertilio murinoides* (later synonymized with *Myotis blythii*) from Chamba area of the state (erstwhile Punjab). Dodsworth (1913 c) recorded seven species of bats from Shimla and adjoining regions. Thomas (1915) reported *Myotis formosus* from Dharamsala and *M. blythii* from Shimla. Besides these, a few occasional species records from the state also exist like *Myotis blythii* (Thomas, 1915), *Scotoecus pallidus* (Allen, 1908), and *Plecotus auritus* (Bhat *et al.*, 1983). 21 mammalian species belonging to 19 genera and 9 orders from Simbalbara wildlife sanctuary were described by Sharma and Saikia (2009). 33 species of mammals are reported in Ladakh, which constitute about 10.7% of Indian mammals **(Mehta, 2009)**. Occurrence of 28 species of bats under 14 genera and five families has been confirmed from Himachal Pradesh by Saikia *et al.* (2011).

Table 2.1: Status of Faunal Diversity in world, India and Himachal Pradesh An Estimation)

Taxonomic group	*No. of Species*		
	World	*India*	*Himachal*
Protista			
Protozoa	31250	2577	89
Animalia			
Porifera	4562	486	01
Cnidaria	9916	486	01
Bryozoa	4000	200	01
Nematoda	30000	2850	127
Annelida			
Oligochaeta	12700	473	46
Hirudinea		59	04
Arthropoda			
Crustacea	35534	2934	04
Arachnida	73440	5818	
Areneae	13	Mites (Soil & Parasitic)	94
Insecta		867391	59353
Odonata	6000	499	77
Lepidoptera (Butterflies)	142500	15000	268
Lepidoptera (Moths)			184
Hemiptera	80000	6500	382
Hymenoptera	120000	10000	319
Dermaptera	2000	320	30
Coleoptera	350000	15500	187
Plecoptera	2100	113	20
Mantodea	2310	162	07
Orthoptera	17250	1750	50
Diptera	100000	6093	14
Ephemeroptera	2200	106	06
Chordata			
Pisces	21723	2546	104
Amphibia	5150	209	17
Reptilia	5817	456	14
Aves	9026	1232	657
Mammalia	4620	390	129

(Sources: Alfred et al., 1998; Anonymous, 2000 a, b, 2005; Mehta and Julka, 2002)

Chapter 3

Khajjiar Lake

Khajjiar lake "The Mini Switzerland of Himachal Pradesh" is present in the western part of Chamba district of Himachal Pradesh (Plate 1). Khajjiar Lake has a clump of reeds and grasses exaggeratedly called an island in it. Fed by slim streams this small lake rests in the centre of large glade of Khajjiar. This glade is greenish in its turf and contains in its centre a small lake having approximate area of 5000 sq. yards. The glade is surrounded from all sides by a thick forest of deodar (*Cedrus deodara*), fir (*Abies pindrow*), and spruce (*Picea smithiana*) (Plate 2).

Physical Features

Fed by tiny streams, this small lake lies at in the centre of large Khajjiar glade. The glade and the lake are held sacred to Khajjinag- after whom the place is named. Dense conifer and broad leaved forests cover the steep mountain slopes around this lake. This lake remains full of water in all seasons. It requires no rain water for survival. For a close view, it has been made accessible with the help of a wooden bridge. A tiny island covered with reeds, keep floating due to divine reasons.

Location

At an altitude of 1983 metres in district Chamba 16 k.m. from Dalhousue and 25 k.m. from Chamba. Khajjiar has thick forest of the Kalatope sanctuary surrounding its soft green grass. In the centre of a grassy meadow' it is 1.5 km long and 1 km wide, and surrounded by cedar forest, a small lake, called the Kund, that forms the centre piece of Khajjiar. Khajjiar is known as 'Mini Switzerland'. The average depth of the lake is sitated to be thirteen feet as per District Gazetteer.

Climate

The climate of Khajjiar in general is tempreate with well defined seasons. However, there may be variations because of micro-climatic systems depending upon altitude and mountain aspect. The winters last from December to February. March and April generally remain cool and dry but snowfall does occur at higher elevations during these months.

The temperature begins to rise rapidly from the middle of April till last week of June or first week of July when monsoon breaks-in. Monsoon continues till the end of August or mid September. During the monsoon, the weather remains misty, humid and cloudy. October and November are comparatively dry but cold.

Historical and Cultural Background

This is the magical paradise called 'Khajjiar Lake'. A temple dedicated to 'Khajjiar Nag' is also located there. It lies in a depression formed by ancient glaciations. It was christened mini Switzerland by Swiss Envoy Willy P. Blazer on 7th July 1992, and was put on world map. In the presence of Indian officials, P. Blazer put up a yellow Swiss hiking footpath sign toward which formally and officially declared Khajjiar as 'mini Switzerland'. The sign board also indicates the actual distance from the Swiss capital Berne upto Khajjiar as 6194. Blazer as per tradition of his country had taken a stone from Khajjiar which was made a part of stone sculpture installed opposite to the Parliament mansion in Berne. Places allover the world similar to Switzerland in respect of geographical and topographical traits and scenic elegance are named after it. Hence, Khajjiar became the 160th tourist spot in the world to be christened as the mini Switzerland.

The Architecture of the original wooden temple of Khaji Naag dated of back to the period earlier to 12th Century A.D. In the 16th Century A.D. Raja Bal Balbhasra Barman erected wooden Panowas ststurs. This temple is renovated by Batlu, the religious nurse of Raja Prithvi Singh in 17th century A.D. In the mandapa of the temple one can see the images of the Pandavas and the defeated Kaurvas hanging from the roof of the circumambulatory path. For the local people this lake holds sacredness and they believe that it is unfathomable. The lake takes its name from Khajji Nag, the deity in the near by temple.

Local Human Population

A Large number of local people reside in villages nearby the lake. Literacy rate is poor as they aremainly involved in vegetable farming and in rearing sheep and goats. Males are involved in horse rearing for horse riding around the lake.

Visitor and Visitor Facilities

Thousands of tourists visit this place every year. Khajjiar is often referred to as "Gulmarg of Himachal Pradesh". The lush green meadows are surrounded by thick pine and cedar forests. Grazing herds of sheep, goats and other milch cattle present a prefect pastoral scenery. There is a small lake in the center of the saucer shaped meadow which has in it a floating island. Much of the lake has degenerated into slush because of heavy silting during rains. Still the landscape of Khajjiar is picturesque and a photographer's delight. There is a Tourism Hotel and some Tourism cottages at Khajjiar where the tourists can stay. Besides, there are two rest houses one each of P.W.D. and Forest Deptt. A couple of private hotels have also come up, which do not match the above places in terms of location and amenities. Bus service to and from Khajjiar is limited and timings change according to local demands. There used to be a golf course in Khajjiar which spoiled over time due to negligence in its maintaintence.

The best entertainment in Khajjiar is to walk around the lake or to go for long walk in the thick pine forests. Children enjoy this place because of the freedom of movement and the slopy terrain which permits them to roll down to the lake without getting hurt. Another attraction like any other hill station is horse riding.

Scientific Research and Activities

Surveyors and geologists from the Himachal Pradesh geological wing surveyed the lake in 1988. Dr N.N. Angires, an agronomist from the Himachal Pradesh Krishi Vishwavidyalaya (HPKV), Palampur, undertook a weed survey in the lake in 1989. He identified 10 species of weeds present in the lake. A block of thalia and salvina, earlier present on an island in the lake that used to float and drift, has stopped due to siltation. The National Institute of Hydrology did a preliminary tour of the lake in 1993 and proposed to undertake battery of tests but the study did not mature.

Conservation value

Lake and its surrounding land are constantly under threat from weeds. Increasing Pressure of solid waste residues is left out by visitors. Large number of tourists comes from Punjab every weekend to this site. They leave solid waste materials and also cause sound and vehicular pollution all over the area. The water quality of the lake is also detourious due to eutrification. Large numbers of animal herds are grazing over the year in this ground which causes water pollution. Horse riding throughout the year is also a contributing factor for water pollution.

Conservation Management

The Khajjiar Eco-tourism Society with the Conservator of Forests (Wildlife), North Zone, Dharamsala, as its Chairman was established in 2002 but no pragmatic solution to resolve the problem has so far been found. The lake, the centre of attraction, has to be cleared of weeds and silt. Wet dredging is the answer, say experts. A horse-riding circular road needs a drain to take the rain water out of the "bowl". The gullies need to be filled. The grazing rights of the local people have to be curtailed. For this, they will have to be adequately compensated and given alternative pastures. According to official sources, to curb unwarranted and unruly activities in the ambients of Khajjiar lake, this beauty spot has been designated as "a special development area" under the Town and Country Planning Act and to look after the cleanliness and development of Khajjiar, a body called, "the Khajjiar Development Board" has been constituted of members of various departments under the chairmanship of the Deputy Commissioner of Chamba. No management plan is prepared or approved by Government. Protection of this lake presents little difficulty given the high altitude, provided that adequate manpower is made available. Despite various studies and solutions worked out by experts in the past two decades, little has been done for the upkeep and beautification of the famous Khajjiar lake here. An effort was made in the early '80s to desilt the lake after pumping out its water. The project registered only a moderate success since organic matter kept choking the pumping equipment. Only some silt from the periphery could be removed. In view of the conjectures that water stagnation in the lake is due to the impermeability developed by the filling up of joints and cracks in igneous rocks of this area, they suggested that upper-forested slopes be channelised, the norther slopes be terraced and protected by a retaining wall and the hydrophytes in the lake be manually removed. Mechanical cutting of weeds below surface and removal, draining out of dirty water, construction of a pucca drainage channel around the outer periphery of the lake and a small stonewall to avoid the entry of material and a physiographic map of the lake by the Geological Survey of India were among the suggestions made. Chemical control of weeds by standardisation of dose etc was also suggested. A proposal for relocation and extinguishing of the grazing rights of people was prepared in 1993. Two courses of action were suggested — to close the area to graziers for 30 years provided that an alternative grazing site is earmarked and, secondly, to acquire rights of the people while providing an alternative common pasture. The closest, viable alternative grazing site was found in the Jhurdu forests 6 km away. It was proposed to divert the main water stream, desilt the lake and provide drainage on the main road, formation of a small earthen bund around the periphery to divert the water from nullah and meadow catchment. For the overall development of this area, it is envisaged to de-weed the lake, divert the water of small nullahs feeding the lake, control silt and develop the meadows from the tourists' point of view. It is proposed to fence the lake so that soil disturbance in its immediate periphery is halted. Manual removal of floating islets of vegetation in the lake is also mooted.

Table 3.1: Water quality of lake

Sr. No.	*Parameters*	*Values*
1	pH	7.45
2	Electrical Conductivity µmhocm-1	
3	Total suspended solids, mg/l	
4	Total alkalinity mg/l	150
5	Dissolve oxygen (DO) mg/l	3.6
6	BOD, mg/l	1.9
7	COD, mg/l	4.75
8	Turbidity NTU	153.9
9	NH3-M	0.051
10	Silicates	93.3

The floristic composition of this area varies from chil pine (*Pinus roxburghii*) with a mixture of ban oak in the lower zone to pure deodar in the middle reaches with culminates in mixed crop of deodar (*Cedrus deodara*) , fir (*Abies pindrow*), and spruce (*Picea smithiana*) species with some alpine pasture towards Dainkund area. 101 species belonging to 54 families and 95 genera occur in Khajjiar beat (Verma and Kapoor, 2011). The dominant families are Asteraceae, Rosaceae, Polygonaceae, Lamiaceae, the number of tress species was 6 and 11 with dominance of *Persea duthiei*, and *Picea smithiana* at 2000m-2200 and 2200m-2400m elevation ranges respectively. The number of shrub species was 9 and 22 with dominance of *Sarcococca sainga* and *Viburnum erubescens* at 2000m-2200 and 2200m-2400m elevation ranges, respectively.

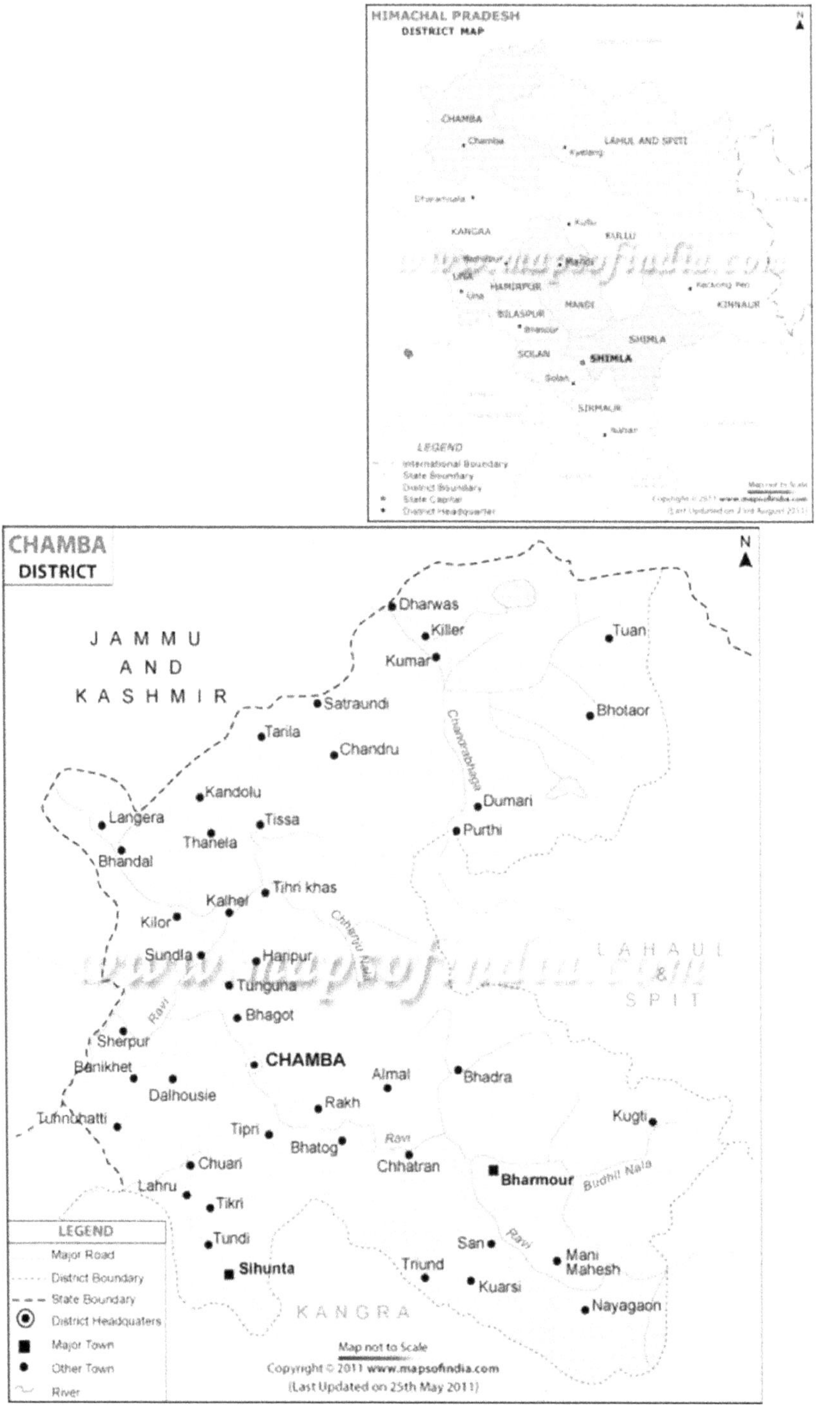

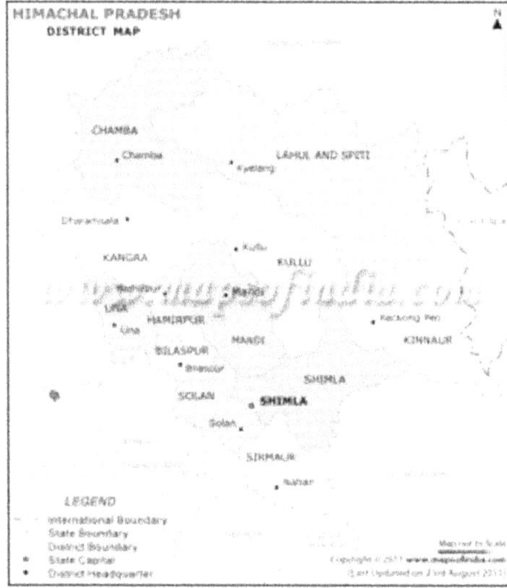

Plate 3.1: Map of Chamba district showing the study area of Khajjiar lake

(Source: www.mapsofindia.com)

The number of herb species was 60 and 54 with dominance of Polygonum capitata and Erigeron alpines at 2000m-2200 and 2200m-2400m elevation ranges, respectively. Undergrowth in the forest is well developed and dense in places with a good cover of grass in November. Pastures land for grazing, number of wild and medicinal plants are found to be here.

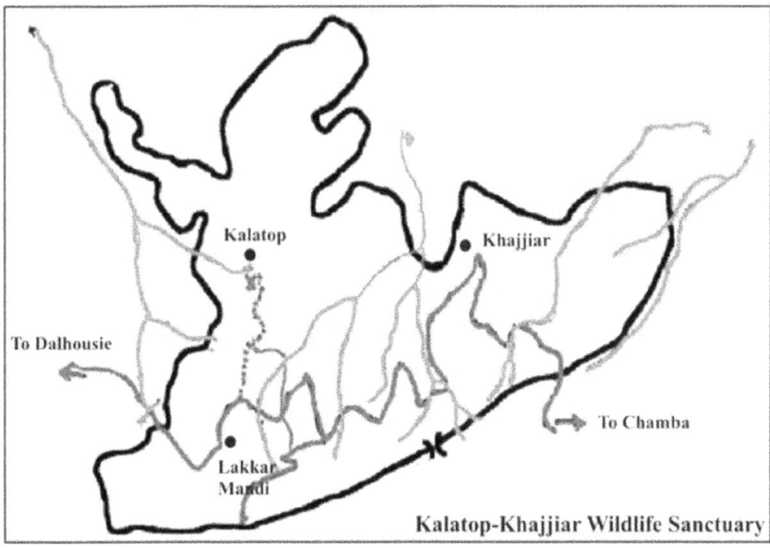

Plate 3. 1 A: Line map of Kalatop-Khajjiar wildlife sanctuary

Plate 3. 1 B: Google imagery of Khajjiar area (Source: www.googleearth.com)

Plate 3.2 A: Khajjiar lake area in winters; B: Khajjiar lake area in summers; C: Deodar forest around lake area

Chapter 4

Methodology to Study Biodiversity

The Khajjiar lake or popularly known as "The Mini Switzerland of Himachal Pradesh" is present in the western part of Chamba district of Himachal Pradesh. Khajjiar Lake has a clump of reeds and grasses exaggeratedly called an island in it. This glade is greenish in its turf and contains in its centre a small lake having approximate area of 5000 sq. yards. The Khajjiar Lake lies 32° 26′ north and 76° 32′east about 6300 feet (1920 meters) above sea level between Chamba and Dalhousie. Khajjiar Lake is situated in Khajjiar- Kalatop wild life sanctuary. This small sanctuary lies in the catchments of the Ravi river, located in western part of Chamba District. It is one of the oldest preserved forests of state (notified on 01.07.1949). Total area of sanctuary is 2,026.89 hectares (20.69 sq. km.) Its mean annual rainfall is 800 mm. Temperature varies from -10° C to 35°C. The climate of Khajjiar varies from being mild in summers and cold and severe in winters. It experiences south-western monsoon rains in July-September. The vegetation consists of mature mixed Blue Pine and Deodar forests, with some Green Oak and Tree Rhododendron.

Study Design

The lower animal groups can be collected, preserved, identified and studied strictly following the wildlife conservation provisions. However, direct observations can be made on large animals in their natural habitats and no individual should be caught or hurt during the observations. Different methods can be employed for study of different animal groups.

Equipment Required

Insect and Planktonic Nets: Nylon net bag of length 75 cm. and mouth diameter 38 cm. on metal ring attached with sturdy and light weight handle of varying size are employed. Planktonic nets of fine musclin cloth can also be used to collect the aquatic faunal diversity.

Aspirator: Aspiratorsare used for collecting small insects and ants. It is made up of two holes in cork stopper of glass jar. Bent glass tube is inserted at one hole to act as intake tube and inverted U shaped glass tube is inserted in another hole whose mouth is closed with a fine piece of cloth. A rubber tube is attached to another end of inverted glass tube for sucking to act as suction tube.

Killing bottles: Before preserving insects, their killing is necessary for which killing bottle is employed. These are prepared with glass jars of wide mouth and tight lid supplied with a layer of Sodium or Potassium cyanide covered with plaster of Parris. Other liquids like chloroform, ethyl acetate are also used. Liquid chemicals are poured over cotton, placed in air tight jar and covered with 3-4 layers of filter papers so that insect specimen remains dry. A killing bottle with a layer of saw dust, soaked with a few drops of ethyl acetate, yields good result and specimens remain fresh and flexible for stretching.

Collection Vials: Killed specimens are preserved in liquid kept in cultural vial of varying size depending upon the size of specimen. These are numbered to note field details correctly.

Paper Packets: Insect specimens cannot be kept in killing bottle for long time as identifying features may be damaged. So after killing, insect specimens are transferred to paper packers for temporary storing and transportation.

Camera: Nikon D 80 fitted with zoom telly lenses can be used for photography. Other cameras with long range lenses can also be employed.

Field note book: It is used for writing details regarding date, place of collection, habitat, colour of live specimen, host plant or host animal, associated insects/animals and other necessary information.

Study of Invertebrates

Lower Aquatic Fauna: Lower aquatic fauna in the lake can be collected in the bottles alongwith water from different sites of the lake regularly. The pH should be determined. The fauna is identified and stained slides are prepared. The protozoan and annelid fauna is preserved according to the manuals.

Arthropods: Due to their immense diversity, documentation of arthropod fauna of an area is never complete, although a good deal of information can be generated on them by planned and systematic surveys. Following methods are used for collection of arthropod fauna:

Hand Picking

Most of insects remain little inactive in morning hours due to excessive cold night, hence, these are picked up by using forceps beneath grasses or below stones. For soft bodied insects, camel hair brush dipped in preserving medium is used to minimize damage to soft body covering of insect. Scissors are used to chop twigs below on which insects take shelter (Jonathan, 1990). Flatworms, annelids, centepedes and some arachnids are picked by forceps. Proper care should be taken during picking up the specimens to avoid any damage to their body parts.

Sweeping

Nylon net with long handle is used for sweeping the insects from herbage. The net used for sweeping consists of a bag, made up of a cloth 50 cm deep, suspended on a 60 cm handle made up of stainless steel. Sweeping is most common method for rapid collection of free flying and free living insects (Arora, 1990).

Aerial Netting

Free flying hymenopterans, lepidopterans and some dipterans are collected by aerial nets. Net with wide mouth is moved in air with the help of long sturdy handle and as insect enters the net, the handle is twisted rapidly. Soft-bodied insects like butterflies may be gently removed from the bottom of the bag, after they get trapped in the bag by a rapid twist of handle.

Aspirator

Aspirator method is used to collect small active insects like leafhoppers, white flies, other Hemiptera and Coleoptera etc. straight from the plant surface. It is also useful to transfer small insects from sweeping nets. The air is sucked by the rubber tubing, which draws the insects to the main tube through the glass tube. The lid of the main tube is removed and the entire content is transferred to the killing bottle or into the preservative.

Light Trap Method

Many insects like moths, beetles etc. are collected by using artificial light on white cloth sheet. These insects are picked up by hand or forceps and killed by using killing bottle.

Preserving Methods

Preservation is very important step in taxonomy. For identification purpose, insect after preservation must look as natural or like alive one. Insect species of different orders have different body make up viz. different colour, soft or hard bodies etc. so they need specific preserving methods.

In many cases after collection, insects are put into killing bottles containing agents like chloroform, ethyl acetate or benzene. These insects are transferred to paper envelope. Each envelopeis numbered carefully and details of specimen number, date, host etc. are written in field notebook. These packets containing specimens are kept in boxes with thin layer of cotton. Thereafter, insects are properly stretched and pinned by using rust free entomological pins of different size depending upon the size of specimens (Arora, 1990). These stretched and pinned specimens are kept in wooden insect boxes in dry conditions. To protect them from fungal infection and other attacks, naphthalene balls are placed in boxes. These balls should be replaced regularly due to their volatile nature. Small insects viz. black ants are mounted on small triangular cards with water soluble adhesive (Jonathan, 1990). Soft-bodied insects of Thysanura, Diplura, Collembola, Protura and many more are preserved in 70% alcohol and glycerol in 9:1ratio.

Study of Vertebrates

Fish, Amphibians and Reptiles

Fish fauna of the lake is trapped in the net and photographed for identification. Living specimens after photography should be transferred back to the lake. No specimen should be brought to laboratory without the permission of the competent.

The amphibians are sampled using the methods described by Vasudevan *et al.* (2001). Amphibians are sampled on the basis of combination of adaptive cluster sampling, visual encounter surveys, audio surveys and opportunistic records. Adaptive sampling is done along the lake and around forest floor. Areas (not Quadrats) of adequate size are selected for study. These areas are searched carefully turning the leaf litter, rocks, as well as by prodding the cavities on the forest floor to look for amphibian species.

The reptiles are best sampled using the general methods described by Ishwar *et al.* (2001). Reptiles are studied by using combination of adaptive cluster sampling and forest transects.

Birds

Stratified Random Sampling Technique (Snedecore and Cochran, 1993) is generally followed for studying the birds of study area. These strategies are mainly based upon the principle of exploration of a portion of the individuals in the whole population. This technique not only allows the collection of the right type of scientific information but also to save time and yield the avian data which is very much amenable for mathematical/statistical analysis for the better presentation of results (Javed, 1996). The other important aspect like the activity of birds should be kept in consideration. Since the peak activity in most birds lasts for 1 or 2 hours after sunrise or before sunset, so monitoring is done either in the early morning or late evening hours.

Birds can be observed with aid of 10x40 Nikon field binoculars. Field identifications can be carried out with the help of various field guides (Ali and Ripley 1983 b; Grimmett *et al.*, 1999; Kazmierczak, 2000). The nomenclature followed here is after Manakadan and Pittie (2001). Residential status of the birds can be worked out and different status categories like resident, winter visitor and summer visitor have been assigned strictly with reference to the study area on the basis of presence or absence method. The birds that show the irregular trend of sighting and population fluctuations (non-seasonal) have been placed under resident with local movements (R/LM) category.

The data recorded in each survey from different areas around the Khajjiar Lakeshould be kept separate and analysis for relative abundance on the basis of frequency of sightings (Mac Kinnon and Philips, 1993). Based upon these, different categories assigned are: Very Common (recorded in more than 45% of data sheets), Common (between 25-45% of data sheets), Uncommon (between 10-24% of data sheets) and Rare (recorded once or twice). The relative frequency scale is fixed in such a way so as to include the migrant species sighted seasonally in good numbers (which visited the area for a brief period of time) to their respective category.

Mammals

The mammalian populations are sampled by using a combination of direct and indirect methods. The direct methods utilize sighting of animals as the main data whereas indirect methods rely on quantification of indirect evidences such as pellet groups, scats, pug marks and hoof marks in a predetermined sampling unit. The mammals can be separated into two main groups based on size i.e. large and small mammals since sampling strategies for both groups differ considerably. The direct evidences of all large and medium sized mammals can be made by using line transects method (Burnham *et al.*, 1980). The entire procedure of line transect sampling is performed by walking on local footpaths due to difficult terrain of the study area. The footpaths are monitored in morning and evening hours which generally coincide with maximum activity period of animals. Some of the transects, where chances of sighting of nocturnal mammals like bats and carnivores is more, can be explored during late evening and night hours. Transects can be walked and monitored with the help of team of local villagers who can scan either side of transect to detect animals.

The indirect evidences such as scats, pellet groups are also employed to study the presence of some mammals. All indirect evidences such as pellet groups, scats are quantified as to species and their number. Different groups of shepherd, local people and Forest Department's employees can also be approached to know the presence of different animals.

The small mammal communities comprising of rats, mice, shrew, squirrels, bats etc. are one of the most vital groups as far as management of any landscape unit is concerned. Data on abundance, distribution and diversity of these groups is collected by general trapping. Sherman traps are deployed in different forest compartments for capturing the rodent species. The traps are set in either morning hours and checked in evening or they are deployed in evening and checked the following morning.

Identifications

Identification of all the species is based on morphological characters. The scientists and experts of related animal group of different research institutes are consulted to confirm the exact taxonomic identifications. Related research literature, identifying keys and earlier records of related species along with their distributional ranges can be taken into consideration. Collected specimens of invertebrates and photographs of vertebrates are used for identification purposes. The identification can be confirmed and authenticated at High Altitude Zoological Regional Station, ZSI, Solan Himachal Pradesh, Northern Regional centre ZSI, Dehradun (Uttarakhand), Wildlife Institute of India, Dehradun, ZSI, Kolkata, Department of Zoology, Punjabi University Patiala Punjab and CSK HPKV Palampur.

Chapter 5

Fauna of Khajjiar Lake

In Khajjiar area of Himachal Pradesh 223 species of different faunal groups (123 invertebrates and 100 vertebrates), spread over 193 genera, 79 families and 32 orders are present (Table 5.3; Fig. 1). Analyses of data shows that class Aves dominate the fauna of Khajjiar with 77 species, followed by Lepidoptera (49 species), Orthoptera (29), Mammalia (16), Coleoptera (15), Odonata (10), Hymenoptera (7), Hemiptera and Diptera (5 each), Reptilia (4) and Amphibia (2 species) (Figs. 2 and 3). It is further analysed that Mollusca, Oligochaeta, Homoptera and Pisces are least represented groups of fauna with a single species each in Khajjiar area.

Faunal diversity of Khajjiar is described here in the following section:

A) INVERTEBRATE FAUNA

A total of 123 species of invertebrates belonging to 110 genera spread over 30 families and 10 orders are found in Khajjiar area. Of these, Lepidoptera (49 species) is the most dominant invertebrate order in the present area, followed by Orthoptera (29 species), Coleoptera (15 species), Odonata (10), Hymenoptera (7), Hemiptera and Diptera (5 each), and Mollusca, Oligochaeta and Homoptera (one species each) (Fig. 3).

I. ANNELIDA: OLIGOCHAETA

Only of a single species of Oligochaeta is there in Khajjiar area of Himachal Pradesh. Very good population of this natural decomposer can be seenin the area. It is further observed that this species somehow passes the extreme winters of Khajjiar in some dormant stage.

Family: Lumbricidae

Normal setae S-shaped, usually eight per segment, seldom more. Spermathecal pores never in segment 4/5, partly at least further back, if not altogether wanting. Clitellar epidermis never of only one layer of epithelium. Testes and funnels in x and xi or one of these.

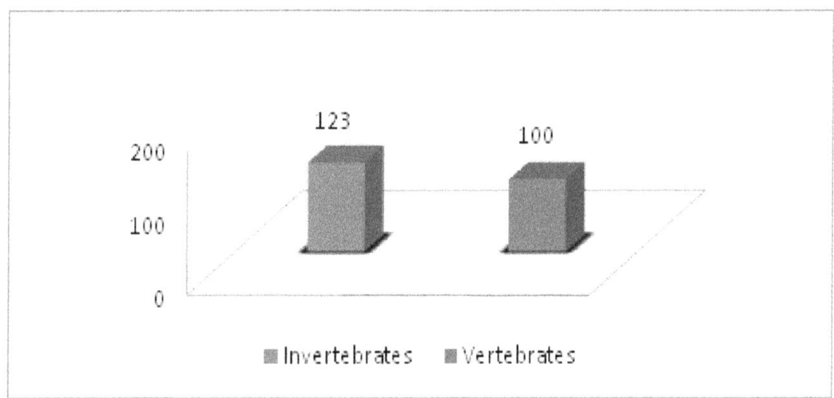

Figure. 5.1: Number of invertebrate and vertebrate species present in Khajjiar area

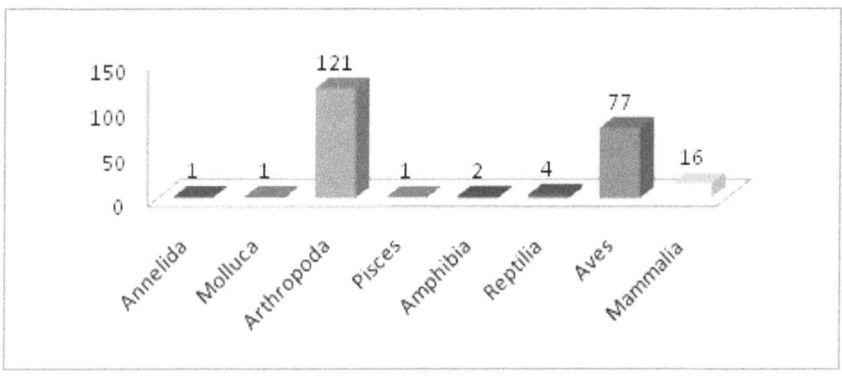

Figure 5. 2: Number of species of different invertebrate and vertebrategroups in Khajjiar area

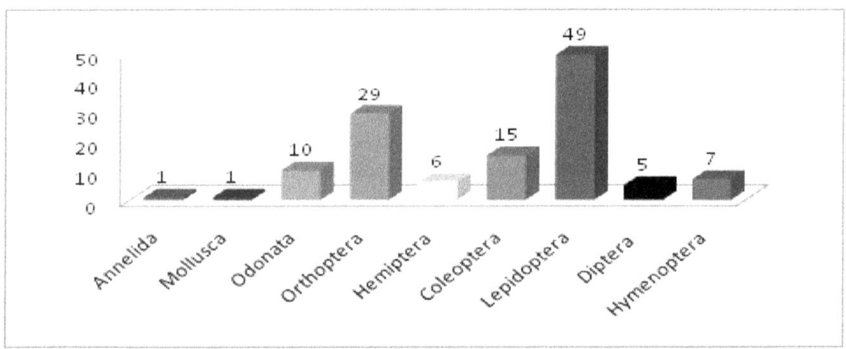

Figure 5. 3: Number of species of invertebrate groups present in Khajjiar area

Genus: ***Lumbricus*** **Linnaeus,** 1758

1758. *Lumbricus* Linnaeus, *Syst. Nat.* 10 ed. 1: 824 pp

Species of this genus usually darkly pigmented. Prostates usually absent (if present, spermathecal pores in groups of several or of several pairs behind testis segments). Three pairs of seminal vesicles present.

1. *Lumbricus castaneus* Savigny 1826 (Chestnut Worm)

(Plate 4 A and B)

1826. *Lumbricus castaneus* Savigny, *Memoires de l Academie des Sciences* 5: 176-184

Diagnostic Characters: Small sized species, usually about 30 to 70 mm and have around 82 to 100 segments. Body shape cylindrical and dorsoventrally-flattened posteriorly. Colour of upper surface of the body red chestnut to violet brown; pale brownish yellow below. A raised whitish gland visible on the underside between the saddle and the head on segment 10. Ventral and lateral setae closely paired. Testis sacs fused to form a single median chamber.

Distribution: An introduced European species of earthworm, found in countries like Mexico, New Zealand, Iceland and United States, also present in Indian subcontinent. In Himachal Pradesh reported from Chamba District only.

Habits and Habitat: A common and widespread species. Found in decaying leaves and manures, in soils with high organic matter content and under logs or stones. Normally present in whole of the meadow area of Khajjiar.

II. ARTHROPODA

Bilaterally symmetrical animals with segmented body; segments usually grouped into two or three regions and paired segmented appendages. Members of this phylum have a chitinous exoskeleton. Exoskeleton periodically shed and renewed as the animal grows. Alimentary canal tubular in shape, starts from mouth and ends at anus. Body cavity hemocoele. Respiration by means of gills or tracheae and spiracles. Excretion by means of tubes that empty into the alimentary canal. Sexes separate.

CLASS: INSECTA

Animals of this class commonly known as insects have three distinct regions of the body i.e. head, thorax and abdomen. Head has one pair of antennae, one pair of mandibles, one pair of maxillae, a hypo-pharynx and a labium. Thorax region has three pairs of legs, one on each thoracic segment. Often two or three pairs of wings present, borne by the second or third of the three segments. Sometimes second and third segments jointly bear the wings. The gonophores present at the posterior end of the abdomen. Abdominal appendages, if present, located at the apex of the abdomen and consist of a pair of cerci, an epiproct and a pair of paraproct. Class insects divided into various orders on the basis of structure of wings and mouth parts.

A total of 121 species of insects belonging to 108 genera spread over 28 families and 8 orders have been recorded from Khajjiar area (Table 5.3). Of these, Lepidoptera (49 species) is the most dominant insect order in Khajjiar, followed by Orthoptera (29

Plate 5. 4: Annelids and molluscs recorded in Khajjiar area

species), Coleoptera (15 species), Odonata (10), Hymenoptera (7), Hemiptera and Diptera (5 each), and Homoptera (1 species).

a) ORDER: ODONATA

Large insects, with beautifully coloured body, spend most part of their life on wings. Immature stages aquatic and adults live around water bodies. All stages of life predaceous, easily recognised. Two pairs of elongated, membranous and veined wings present. Compound eyes large and many faceted, occupy most of the head. Thorax relatively small and compact. Prothorax always small and other two thoracic segments make most of the thorax. The antennae very small and bristle like. Abdomen slender and long. Cerci unsegmented and functions as clasping organ in males. Moth parts chewing type and metamorphosis simple.

10 species belonging to 8 genera spread over 5 families of odonates are there in Khajjiar lake area. Based on morphology, the members of order Odonata belonging to 2 groups, *viz*. damselflies (Zygoptera) and dragonflies (Anisoptera) are recorded from the khajjiar. It is analysed that family Libellulidae supports the highest number of species (6 species, under 4 genera) and all other families are represented by a single species each.

Key to the Groups (Sub-Orders)

1. Hind wings broader than the forewings; abdomen stout; the wings held perpendicular to the body...........Dragonflies (Anisoptera)
2. Fore and hind wings narrowed at base; similar in size and shape; abdomen slender; usually the wings kept closed over the body.. Damselflies (Zygoptera)

Sub-Order: Anisoptera

Family: Cordulegasteridae

Mountain Hawks-the dragonflies of very large size, some being the largest of the order Odonata having black or dark brown colour dragonflies with bright yellow markings. Head robust and transversely elongated. Eyes large and moderately separated or meet at a point. Prothorax very small, entirely hidden by head. Thorax robust. Wings transparent or tinted with golden yellow, the hind usually broader than the fore. Abdomen cylindrical in both sexes or compressed in females.

Genus: *Anotogaster* Selys, 1854

1854. *Anotogaster* Selys, ***Bull. Acad. Belg.* (2) vol. xxi, p. 101**

Head very massive, eyes meeting at one point only, slightly tumid behind. Face deeper than broad, not concealing or overlapping the eyes in any way. Antennae seven-jointed, the basal joint short, rounded, the second long and very robust, the third as long as second but slim, the fourth to sixth each less than half the length of third, the terminal or seventh very short and filamentous. Prothorax short and massive, posterior lobe rounded, tumid. Thorax relatively massive usually coated with fine downy hairs, especially on dorsum.

1. *Anotogaster basalis* Selys

(Plate 5A)

1854. *Anotogaster basalis* Selys, *Bull. Acad. Belg*. (2) vol. xxi, p. 102

Size: Male: Abdomen: 53-56 mm, Hind wing: 42-44 mm.

Female: Abdomen: 59 mm, Hind wing: 51mm.

Diagnostic Characters: *Male,* frons in front bright citron-yellow narrowly bordered below with black, its upper surface citron-yellow, the basal half black, this colour extending to the sides. *Pro-thorax* black with a basal ring and the border of posterior lobe narrowly yellow. *Thorax* black, marked with greenish-yellow. Two pyriform antehumeral stripes, very broad and in close apposition above, tapered to a fine point. Laterally two broad oblique stripes. *Legs* black in colour. *Wings* hyaline. Abdomen black, broadly ringed with citron yellow. *Female,* very similar to the male in colour and markings. *Abdomen* more robust, markedly laterally compressed; segment 9 aborted, oblique, and produced ventral wards into a long ovipositor which extends well beyond end of abdomen. Adult females have the face-markings more restricted than in the male. Abdominal markings broader than in the male.

Distribution: Found in north western India, reported from Punjab, Kumaon hills and Bengal.

Habits and Habitat: A dragonfly of fresh water ponds. Found up to 4,600 feet. Present in good numbers from Khajjiar lake.

Family: Aeshnidae

Dragonflies of this family large or medium sized, homogenous in shape, variable in colour and markings. Eyes meet broadly over head. The wings transparent, long, moderately broad, base of hind wing in males usually more or less excavate, rarely rounded at tornus in both sexes. The abdomen longer than the wings and often tumid at the base. Most of the darners inhabit marshes, ponds and lakes.

Genus: *Anax* Leach

1815. *Anax* Leach, *Edinb. Encycl*. vol. ix, p. 137

Wings hyaline, but often partly tinted with yellow or pale brown. Head very large and globular. Eyes broadly contiguous. Frons with sharply angled fore-border or crest, but not elevated; occiput very small. Thorax robust; legs long and robust, femora armed with rows of short, closely-set spines.

2. *Anax immaculifrons* Rambur, 1842 (Blue Darner)

(Plate 5B)

1842. *Anax immaculifrons* Rambur, *Ins. Nevrop*. p. 189

Size: Male: Abdomen: 52-55 mm, Hind wing: 55 mm.

Female: Abdomen: 56 mm, Hind wing: 58-60 mm.

Diagnostic Features: *Male,* face pale bluish-green, with a very narrow black border to base of frons above. Eyes sapphire blue, narrowly bordered with black behind. *Prothorax* dark reddish-brown, paler laterally, posterior lobe with heavy fringe of long hairs. Thorax pale bluish green dorsally and turquoise blue laterally with two black stripes. Legs black. Wings transparent, tinted with amber yellow from apex to base of discoidal cell, palely at apex, rather deeply towards base of wing. The first segment of abdomen black and the second segment turquoise blue with a black suture and a middorsal transverse mark shaped like 'Sea gull bird'. Segment 3 with its base laterally broadly turquoise blue. Segments 4 to 8 with apical half black this gradually changing to pale reddish-brown towards base of segments and finally pale dirty blue. The segment 9 black and 10th segment black or brown on dorsum. *Female*, very similar to the male, but the turquoise-blue replaced by pale greenish-yellow on thorax and base of abdomen. Segment 1 warm reddish-brown instead of black; dorsum of thorax pale brown instead of bluish.

Distribution: Found in whole of the Oriental region. In India recorded from Mumbai, eastern Ghats and northern Ghats. Relatively less common in Himalayas.

Habits and Habitat: A common insect at altitudes from 1500 to 7500 feet. Frequents slow flowing streams. Breeds in hill streams. Female inserts eggs into a submerged water plant. Present in good numbers from meadow area of Khajjiar.

Family: Libellulidae

Insects of this family, commonly known as Skimmers-most diverse group of all the odonates. Large, medium or small dragonflies, wings vary in size, shape, width and colouration. Eyes always broadly confluent. Breed in wide variety of aquatic habitats like puddles, ponds, marshes, rivers, domestic storage tanks and aquaria.

Key to the genera

1. More than 12 antenodal nervures in forewing; shape of abdomen variable .... *Orthetrum*
2. Eyes more broadly contiguous; dis-coidal cell in hind-wing traversed; costal border of fore-wing sinuous near base; frons metallic above; discoidal field beginning with at least 3 rows of cells .... *Palpopleura*
3. Discoidal cell in forewing broader, its costal side about one-half the length of basal; no supplementary nervure .... *Trithemis*
4. Wings with small basal yellow markings; face red .... *Crocothemis*

Genus: *Orthetrum* Newman

1833. *Orthetrum* **Newman, *Ent. Mag.* vol. i, p. 511**

3. *Orthetrum Sabina* (Drury) (Green Marsh Hawk)

1770. *Libellula sabina* Drury, *IB. Exot. Ins.* vol. i, pi. xviii, pp. 114, 115

1889. *Orthetrum sabina* Kirby, *Trans. Zool. Soc. Lond.* vol. xii, p. 302

Size: Male: Abdomen: 30-36 mm, Hind wing: 30-36 mm.

Female: Abdomen: 35 mm, Hind wing: 31-35 mm.

Diagnostic Characters: *Male,* face and frons yellowish, becoming brighter citron-yellow on upper surface of the later and variably marked on anterior surface with black or dark brown. Frons very deeply notched so as to form two triangular facets in front. Eyes green mottled with black. *Pro-thorax* bright yellow, with anterior and middle lobes blackish-brown posteriorly. Thorax Greenish yellow with black tiger like stripes. Legs black and inner side of anterior femora yellow. Wings transparent; inner edge of hind wing tinted with yellow. Membrane dark brown. Abdomen greenish-yellow, marked with black. The segments 1-3 green with broad black rings and distinctly swollen at the base. Segments 4 to 6 with a broad oval dorsal black spot on basal third of segments. Segments 7 to 9 black. Segment 10 with base broadly so, apical border finely black. *Female,* very similar to the male both in colour and the remark-able shape of abdomen, differing only in sexual characters. Anal appendages pale yellow, shortly conical.

Distribution: Widely distributed in Ethiopia, Oriental and Australian region. Found throughout the Indian subcontinent.

Habits and Habitat: On wings, throughout year except the months of extreme cold, found upto the altitude of 2000 m. One of the most predaceous of all dragonflies, eats even its own species. A common dragonfly of gardens and fields. Perches motionless on shrubs and dry twigs for a long time. Species can be seen far away from water and occasionally enters houses at night. Hawks flying insects such as flies, small butterflies and dragonflies. Generally, found in good numbers from the Khajjiar meadow.

4. *Orthetrum triangulare* (Selys) (Blue-tailed Forest Hawk)

(Plate 5 C)

1878. *Libella triangularis* Selys, *Mitth. Mus. Dresden,* p. 314

1886. *Orthetrum triangulare* Kirby, *Proc. Zool. Soc. Lond.* p. 327

Size: Male: Abdomen: 29-33 mm, Hind wing: 37-41 mm.

Female: bdomen: 29- 32 mm, Hind wing: 37 mm.

Diagnostic Characters: A medium sized dragonfly with black thorax, black-brown patch at wing bases and blue tail. *Male,* face glossy black, behind head black with a single yellow spot; eyes dark blue; *prothorax* and *thorax* velvety-black; *legs* black; *wings* hyaline, with a broad triangular blackish-brown spot at the base of hind-wing, membrane black. Abdomen broad at base and gradually tapering towards the tip; segments 1-2 and 8-10 black; segment 3-7 azure blue, covered with fine hair. Anal appendages black. *Female,* face dark brown. Mid-dorsum of thorax olivaceous-green often suffused at mid-dorsal carina with reddish-brown. The sides dark reddish brown with two bright yellow stripes. *Wings* more often suffused with brown. The hind wing tinted with yellow lacks basal black area. Abdomen black and without fine hairs. A mid-dorsal yellow or olivaceous green stripe runs from segments 1-7. The segments 2-7 have two yellow spot underneath.

Distribution: Distributed throughout the Oriental region. Occurs all along the Himalayas from Kashmir, Bengal to Myanmar.

Habits and Habitat: Typically hilly species usually found in marshes associated with hill streams. Breeds in brooks flowing through marshes in foothills. Can be seen in good numbers in Khajjiar meadow.

5. *Orthetrum pruinosum neglectum* (Rambur) (Marsh Hawk)

(Plate 5 D)

1842. *Libellula neglecta* **Rambur, *Ins. Nevrop.* p. 86**

1886. *Orthetrum pruinosum* **Kirby, *Proc. Zool. Soe. Lond.* p. 327**

1890. *Orthetrum neglectum* **Kirby, *Cat. Odon.* p. 182**

1931. *Orthetrum pruinosum neglectum* **Fraser, *Rec. Ind. Mus.* vol. xxxiii, p. 446**

Size: Male: Abdomen: 28-31 mm, Hind wing: 32-36 mm.

Female: Abdomen: 25-30 mm, Hind wing: 31-36 mm.

Diagnostic Characters: *Male,* face ochreous to pale reddish brown; eyes blue-black above and bluish-grey below; *pro-thorax* and *thorax* reddish-brown to dull purple in colour; *legs* black and reddish brown at the base. *Wings* transparent, in old adults, pale brown towards the tip; basal area marked with reddish brown in both wings. Wing spots reddish brown. *Abdomen* bright red but in old adults, purplish due to pruniscence; *anal appendages* red. *Female,* differs widely from the male. Face pale olivaceous in colour; *thorax* reddish-brown or dull ochreous, with an ill-defined brown stripe on each side of dorsum. *Wings* similar to male, but the basal marking paler and almost obsolete. *Abdomen* dull ochreous with each segment thinly bordered with black, sides of segment 8. *Anal appendages* dark ochreous, shortly conical.

Distribution: Found throughout India and Myanmar. Also found in Tibet, China and Hing- Kong.

Habits and Habitat: One of the commonest dragonflies in the plains and found around wells, ponds, ditches, tanks and rivers. Males very conspicuous and seen perched on shrubs and stones, seen on wings throughout the year up to 7250 feet. Breeds in puddles, ponds and tanks, also pools in river beds. Present in Khajjiar area mostly during summer months.

Genus: *Palpopleura* Rambur

1842. *Palpopleura* **Rambur, *Ins. Nevrop.*, 26, 129**

6. *Palpopleura sexmaculata* (Fabricius) (Blue-tailed Yellow Skimmer)

(Plate 5 E)

1787. *Libellula sexmaculata* **Fabricius, *Mant. Ins.* vol. i, p. 338**

1868. *Palpopleura sexmaculata* **Brauer, *Verh. zool.-bot. Ges. Wien,* vol. xviii, p. 716**

Size: Male: Abdomen: 14-16 mm, Hind wing: 15-21 mm.

Female: Abdomen: 13-14 mm, Hind wing: 18-21 mm.

Diagnostic Characters: A small dragonfly with greenish yellow thorax and blue abdomen. *Male,* face creamy yellow with brilliant iridescent blue frons; *prothorax* dark brown, with posterior collar and a geminate spot on dorsum of middle lobe bright yellow. *Thorax* pale greenish-yellow, marked with dark brown and black. Dorsal side warm reddish brown. *Legs* bright yellow, flexor surface of tibiae, tarsi, and outer sides of middle and anterior pairs of femora black. Forewings transparent with three black streaks extending from the wing base to the tip. The hind wings tinted with yellow and have two short black streaks extending from the wing base to the tip. Abdomen light blue and covered with pruinescence. *Female,* differs from males in many aspects. Face yellow without iridescent; markings of thorax and pro-thorax very restricted. Thorax rich orange-brown with lateral brown stripe. Wings transparent and more broadly marked with blackish-brown and black and more deeply tinted with amber yellow. Abdomen bright reddish brown with a median black stripe.

Distribution: Found throughout the western India, Tibet, Malayasia and China.

Habits and Habitat: Inhabits ponds marshes, lakes and tanks. Shows slow, circling flight. During summer months, good population of this species is visible in Khajjiar area.

Genus: *Crocothemis* Brauer

1868. *Crocothemis* Brauer, ***Verh. Zool. Bot. Ges. Wien*, vol. xviii, (2): 367, 736**

7. *Crocothemis servilia* (Drury) (Ruddy Marsh Skimmer)

(Plate 5 F)

1770. *Libellula servilia* Drury, ***IIIEx. Ins.* vol. xlviii, pp. 112-113**

1868. *Crocothemis servilia* Brauer, ***Verh. Zool. Bot. Ges. Wien*, vol. xviii, pp 737**

Size: Male: Abdomen: 24-35 mm, Hind wing: 27-38 mm.

Female: Abdomen: 25-32 mm, Hind wing: 31-37 mm.

Diagnostic Characters: Medium sized, blood red or reddish yellow dragonflies with amber coloured patch at wing base. *Male,* face and frons blood red; eyes blood red above, purple on the sides and paler below; prothorax ferruginous, with a spot on middle of anterior lobe; thorax bright ferruginous, often blood red on dorsum. Colour of legs reddish; wings transparent base; abdomen blood red, segment 8 and 9 with mid-dorsal carina blackish. *Female,* differs widely in colouration from male. Face pale yellow; eyes brown above and olivaceous below; prothorax and thorax olivaceous brown, often tinted with ferruginous. Legs dark brown; wings similar to males but basal amber marking paler than in the males. Abdomen yellowish brown with a mid dorsal black stripe.

Distribution: Widely distributed in Oriental and Australian regions. Also present in South Asia, Japan and Philippines.

Habits and Habitat: One of the commonest red dragonflies. Perches on aquatic weeds and chases any passing-by dragonflies. Frequently found in ponds, puddles, rivers, big wells, tanks, ditches and paddy fields. Breed in marshes associated with ponds, rivers and tanks. Present in good numbers in Khajjiar area during summer and monsoon months.

Genus: ***Trithemis*** **Brauer**

1868. *Trithemis* Brauer, ***Verh. Zool. Bot. Ges. Wien***, **vol. xviii, pp. 176**

8. *Trithemis festiva* (Rambur) (Black Stream Glider)

(Plate 5 G)

1842 *Libellula festiva* **Rambur, *Ins. Nevrop.*, pp. 92**

1868. *Trithemis festiva* Brauer, ***Verh. Zool. Bot. Ges. Wien***, **vol. xviii, pp. 736**

Size: Male: Abdomen: 22-28 mm, Hind wing: 26-32 mm.

Female: Abdomen: 21- 24 mm, Hind wing: 29 mm.

Diagnostic Characters: *Male,* frons dark brown in front and iridescent violet above; eyes dark brown above with a purple tinge, bluish grey laterally and beneath. Thorax black coated with purple pruinescence. Legs black; wings transparent, with a dark opaque brown mark at the base of hind wing. Abdomen black and segment 1-3 covered with fine blue pruinescence. Anal appendages black. *Female,* differs markedly from the males. Face dirty brown in front and changes to brown above. Eyes dark brown above and grey below. Thorax greenish yellow to olivaceous with a medial and lateral dark brown stripe. On the sides inverted 'Y' shaped stripes present. Legs black and anterior femora yellow on the inner side. Abdomen cylindrical and of equal length throughout. Colouration bright yellow with medial, lateral and ventral black stripes.

Distribution: Found throughout Oriental region. Also found in Myanmar, Malaysia and China.

Habits and Habitat: Common in slow flowing streams and canals. Usually perches on boulders and aquatic plants. Breeds in still waters or more commonly in sluggish streams. Thses can be seen on wings from May to October in Khajjiar area.

Sub-Order: Zygoptera

Family: Coenagrionidae

Members of this family commonly known as Marsh Darts include slender and small damselflies with varied colouration. These damselflies rest with wings closed over their body. Wings transparent and rounded at the tip and may or may not have wing spot. The long and slender abdomen slightly longer than the hind wing. Though most of the species closely associated with aquatic habitats, some species can be found far away from any aquatic habitat.

Genus: ***Pseudagrioni*, Selys**

1876. *Pseudagrionii*, Selys, ***Bull. Acad. Belg.*** **(2) vol. xxii, p. 490**

9. *Pseudagrioni* sp.

(Plate 5 H)

Size: Abdomen: 34 mm, Hind wing: 22 mm.

Diagnostic Characters: Identified up to genus level based upon characters viz., medium sized and slender built damselfly with non-metallic bright blue colour with black stripes. Head narrow; eyes globalate; prothorax pale blue, posterior lobe bluish green, middle lobe with base narrowly black. Thorax rather slender, azure blue, the mid dorsal with a black band, equal black lines run on both sides of thorax. Legs short, labial spines of moderate length. Wings transparent; pterostigma present but very narrow, braced, covering less than a segment. Abdomen slender, cylindrical, not nearly twice the length of hind wing. Abdomen azure blue in colour with variable black markings from segment 1 to 10.

Distribution: Genus widely distributed to the Old World. Found in Africa, Madagascar, India, Myanmar, Malaysia, Java, Sumatra, Australia, China and Philippines.

Habits and Habitat: Frequents banks of rivers, ponds, lakes and other water bodies. Usually perches on aquatic plants on the bank and usually found in small groups of 3-4 individuals. This spcies can be on wings during summer months in the Khajjiar area.

Family: Calopterygidae

Commonly known as Glories; large damselflies with broad head and conspicuous round eyes. Wings transparent, amber or iridescent coloured; hind wings broad, rounded and iridescent coloured. Size of abdomen longer than the hind wings. Found both in temperate and tropical regions; associated with forested streams and breed in them.

Genus: *Neurobasis* Selys

1853. *Neurobasis* Selys, ***Syn. Cal.*** **p. 17**

Fore wings of male transparent, while hind wings opaque and coloured partly with brilliant metallic green and blue. All wings of female hyaline but with an opaque whitish spot at nodes. Pterostigma absent in males, false and whitish in females. Males commonly seen flitting up and down the stream, hugging the surface of water very closely. Rests on over hanging ferns and herbage besides the stream, or commonly perch with closed wings on a rock.

10. *Neurobasis chinensis chinensis* (Linnaeus, 1758) (Stream Glory)

1758. *Libellula chinensis*, Linnaeus, *Syst. Nat.* Vol. I, p. 515

1890. *Neurobasis chinensis* Kirby, *Cat. Odon.* p. 102

Size: Male: 45-50 mm, Hind wing: 32-38 mm.

Female: Abdomen: 44- 50 mm, Hind wing: 36-40 mm.

Diagnostic Characters: *Male*, antennae with basal and second joint pale blue. Eyes with two sharply defined area viz., upper two third blackish brown and lower third bluish green. Prothorax bronzy green with a coppery reflex. Thorax brilliant metallic green; legs very long and slim. Wings moderately rounded at tips especially hind wings. Forewings transparent, tinted with pale yellowish green with emerald green venation. Hindwings opaque, basal two thirds iridescent green or peacock blue. Abdomen narrow and cylindrical much longer than wings, bronzy green above

A *Anotogaster basali*

B *Anax immaculifrons*

C *Orthetrum triangulare*

D *Orthetrum pr. Neglectum*

Plate 5. 5: Number of species of invertebrate groups present in Khajjiar area

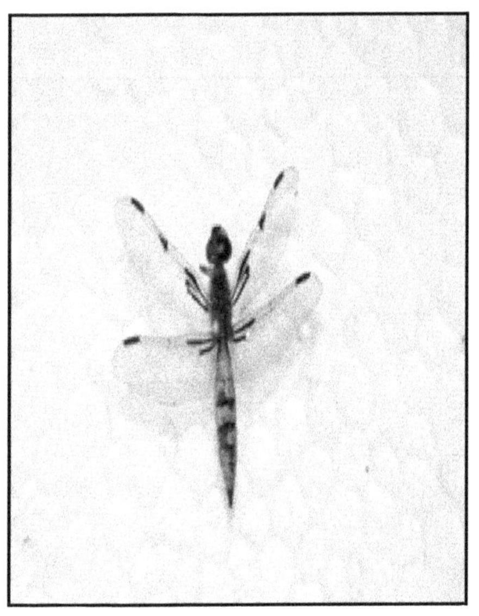

E *Palpopleura sexmaculata*

F *Crocothemis servilia*

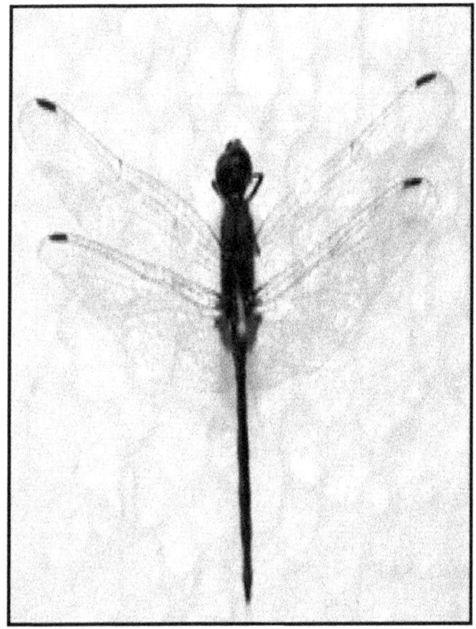

G *Trithemis festiva*

H *Pseudagrioni* sp.

Plate 5.6: Some odonates recorded in Khajjiar area

and on sides, under sides of abdomen black. *Female,* prothorax and thorax as in males, but humeral and lateral sutures finely white, with black borders. Wings tinted with yellow, palely infused with brown, especially at the apices and along the costa in forewing and generally deeper in tint throughout the whole of hind wing.

Distribution: Found throughout India except desert area, from sea level up to an altitude of 2250 m. Commonly found between 500-1200 m.

Habits and Habitat: Common in hill streams. Perches on emergent boulders and fallen logs in streams. Males flash its iridescent green marking of hindwing immediately after alighting. Female lays eggs on submerged decaying logs in streams during southwest monsoon. This species are seen on wings from May to October in Khajjiar area.

B). Order : Orthoptera

Winged or wingless insects, winged forms usually have four wings. Front wings generally elongated, many veined, somewhat thickened and referred to as tegmina. Hind wings membranous, broad and many veined and at rest folded fanwise beneath the forewings. Some species have one or both of the wings greatly reduced or absent. Body elongated and cerci well developed. Antennae relatively long and many segmented. Many species have long ovipositor, in others short and more or less hidden. The tarsi usually three or five segmented. Mouthparts chewing types and metamorphosis simple.

29 species of Orthoptera belonging to 28 genera, spread over 5 families are supportrd by Khajjiar lake area. Analyses of data on orthopterans revealed that family Acrididae supported the maximum number of species (16 species), followed by Tetrigidae (6 species), Gryllidae (4 species), Pyrgomorphidae (2 species) and Conocephalidae (1 species). It has been reported that all the orthopteran genera except *Ergatettix* have been represented by a single species each.

Key to the Suborders

1. Antennae shorter, with less than 30 or fewer segments; tympanal organs, when present, at the base of abdomen

   ....Caelifera

- Antennae about as long as or longer than body, many segmented; tympanal organs, when present, on fore tibiae ....Ensifera

Sub-Order: Caelifera

Key to the Superfamilies

1. Antennae shorter, with less than 30 segments; tympanal organs, when present at base of abdomen 2

- Antennae about as long or longer than body, many segmented; tympanal organs, when present on fore tibiae 3

2. Pronotum normal or if rarely extended behind, then empodium present, and antennae shorter than fore femur..Acridoidea
- Pronotum extended backwards to cover abdomen; empodium absent; antennae longer than fore femur...Tetrigoidea
3. Tarsi 3 segmented .... Grylloidea
- Tarsi 4 segmented, at least on middle and hind legs.................Tettigonioidea

Superfamily: Acridoidea

Key to the families

1. Foveolae lateral or inferior, never forming the tip of the fastigium, stridulatory organs present Acrididae
- Foveolae of the vertex contiguous, superior and forming the extremity of the fastigium, stridulatory mechanism absent

  ....Pyrgomorphidae

Family: Acrididae

Key to the Subfamilies

1. Prosternal tubercle usually absent 2
- Prosternal tubercle present 5
2. Stridulatory serration on inner side of hind femur absent

   ....Acridinae
- Stridulatory serration on inner side of hind femur present

  ....Truxalinae
3. Stridulatory mechanism represented by row of articulated pegs on inner side of hind femur ....Gomphocerinae
4. Head rounded in profile, face almost vertical, rarely oblique and tegmina always with an intercalary vein in median area

   ....Oedipodinae
5. Radial area of tegmina with a series of regular, parallel stridulatory veinlets...Hemiacridinae
- Stridulatory veinlets of radial area of tegmina abse..................6
6. Lower external lobe of hind knee with spine-like apex

   ....Oxyinae
- Lower external lobe of hind knee with apex rounded, angular or subacute, but not spine like 7
7. Last abdominal tergite in male with well developed furcula; supra-anal plate mostly with attenuate or trilobite apex .....Coptacrydinae
- Last abdominal tergite without well developed furcula; supra-anal plate

variable 8

8. Mesosternal lobes rounded or obtuse angular or acute-angular, but not rectangular 9

- Mesosternal lobes rectangular ....Cyrtacanthacridinae

9. Dorsum of pronotum flat or weakly tectiform, with median and lateral carinae linear; male cercus with strongly compressed, lobiform or sub acute ....Eyprepocnemidinae

- Dorsum of pronotum of variable shape; lateral carinae, if present, not linear; male cercus variable, but not as mention above ....Catantopinae

Subfamily: Acridinae

Key to the genera

1. Hind femur, with a spine on the dorso-external and dorso-internal genicular lobes, head small, equal to or less than pronotum

   ....*Gelastorrhinus*

2. Hind femur, with rounded dorso-external and dorso-internal genicular lobes, antennae filiform ....*Ceracris*

Genus: *Ceracris* Walker

1870. *Ceracris* Walker, *Cat. Derm. Salt. B.M.* iv: 790

1. *Ceracris nigricornis nigricornis* Walker

(Plate 6 A and B)

1870. *Ceracris nigricornis* Walker, *Cat. Derm. Salt. B.M.* iv: 791

1925. *Ceracris nigricornis nigricornis* Uvarov, *Ent. Mitt. Berl.* 14: 13

Size: Length: 32 mm.

Diagnostic Characters: Overall colour brown, with scattered black dots. Antennae narrowly ensiform, shorter than the head and pronotum together, brown at the tips. Pronotum truncated in front, and very obtusely angulated behind. Tegmina moderately broader and longer than the abdomen, truncated towards the tip. Wings narrow, bluish hyaline, with greenish nervures.

Distribution: Found in Arunachal Pradesh, Assam, Bihar, Himachal Pradesh, Tamil Nadu, Uttar Pradesh and West Bengal. *Elsewhere,* Afghanistan.

Habits and Habitat: Very common, found throughout the year mostly in grasses and does not show any seasonal variation. This species is present mainly in the meadow area, mainly during monsoons.

Genus: *Gelastorrhinus* Brunner

1903. *Gelastorrhinus* Brunner, *Ann. Mus. Genova,* 33: 137, 157

2. *Gelastorrhinus laticornis* (Serville)

1839. *Opomala laticornis* Serville, *Ins. Orth.*: 590

1914. *Gelastorrhinus laticornis* Kirby, *Fauna British India, Orthopt.*, (Acrid.): 217

Size: Length: 20-22 mm.

Diagnostic Characters: Head large and conical; antennae much thickened and flattened at base; pronotum tricarinate; a dark lateral stripe runs behind each antenna, interrupted by the eye but beyond to the pronotum and below the lateral carinae; tegmina longer than the abdomen, linear, obtusely pointed at tip, uniformly yellowish green.

Distribution: In India found in Maharashtra and West Bengal. Also present in Bangladesh.

Habits and Habitat: Common and abundant in rainy season in grasses. This species is generally recorded from April to October in Khajjiar area, mostly in long grasses.

Subfamily: Gomphocerinae

Genus: *Dnopherula* Karsch

1896. *Dnopherula* Karsch, *Ent. Zeit Stettin.* 57: 259

Fastigium with truncate apex; frontal ridge without lateral carinulae.

3. *Dnopherula (Aulacobothrus) decisus* (Walker)

(Plate 6 C)

1871. *Stenobothrus decisus* Walker, *Cat. Derm. Salt. Brit. Mus.*, 5: 827

1921. *Aulacobothrus decisus* Uvarov, *Ann. Mag. Nat. Hist.*, **7** (9): 482

Size: Length: 14 mm.

Diagnostic Characters: Small sized species with brownish colouration. A pale stripe usually runs starting from fastigium of vertex to ends of pronotum; tegmen testaceous; wings hyaline, with nervures of the costal areas reddish; fastigium of vertex almost subtriangular; abdomen carinated above; tibia with 12 small black spines on each carina.

Distribution: Found in Maharashtra and West Bengal in India, and also present in Sri Lanka.

Habits and Habitat: Very common species which does not show any seasonal variation. Found throughout the year, everywhere, mostly in agriculture fields. This species is present in the grasses around the Khajjiar lake.

Subfamily: Oedipodinae

Key to the Genera

1. Median carina of pronotum entire or slightly intersected by the transverse groove..*Aiolopus*

2. Basal half of tegmina opaque ....*Locusta*
3. Tegmina with the basal third opaque, light reddish brown, very thickly reticulated; the rest subhyaline.........*Pseudosphingonotus*

Genus: *Aiolopus* Fieber

1853. *Aiolopus* Fieber, *Latos*, iii: 100

4. *Aiolopus thalassinus tamulus* (Fabricius, 1798)

1798. *Gryllus tamulus* Fabricius, *Ent. Sys.*, Suppl.: 195

1968. *Aiolopus thalassinus tamulus* Hollis, *Bull. Br. Mus. Nat. Hist.* (Ent.) 22 (7): 347

Size: Length: 17-25 mm

Diagnostic Characters: Fastigium with front angle more acute and foveolae narrowly trapezoid, about twice as long as wide; frontal ridge gradually narrowing and almost angular towards fastigial end. Pronotum somewhat saddle-shaped, posterior margin rounded. Colour of posterior tibiae usually red apical fourth and broadly separated from black band by a wide bluish grey band.

Distribution: Found in Andaman Islands, Bihar, Delhi, Himachal Pradesh, Karnataka, Madhya Pradesh, Rajasthan, Tamil Nadu and West Bengal. Also present in Australia, Bangladesh, Borneo, Celebes, China, Hainan, Hong Kong, Japan, Java, Lombok, Malaya, Myanmar, New Guinea, Papua, Philippines, Singapore, Sri Lanka, Sumatra, Taiwan, Thailand and Timor.

Habits and Habitat: Common and mostly reported in spring and monsoon seasons. Found from February to November mostly in agriculture fields around the lake areas. A serious pest of wheat, millets, tobacco, cabbage, oat, pea and other summer crops. Generally recorded from grassy areas around the Khajjiar Lake and also agricultural fields around the area.

Genus: *Locusta* Linnaeus

1758. *Locusta* Linnaeus, *Syst. Nat.* ed. x, i: 431

5. *Locusta migratoria danica* (Linnaeus, 1767)

1767. *Gryllus Locusta danicus* Linnaeus, *Syst. Nat.* 12 (2): 702

1921. *Locusta migratoria danica* Uvarov, *Bull. Ent. Rec. Lond.*, 12 (2): 137, 161

Size: Length: 32-60 mm.

Diagnostic Characters: Overall colour greenish brown, banded with brown. A brown stripe runs behind the eyes, generally intersected by a white line on the head, but do not extend to the extremity of the pronotum. Tegmina subhyaline, reticulated with brown; wings greenish or yellowish hyaline. Hind femora green, hind tibiae red. Male much smaller than the female.

Distribution: Cosmopolitan, found almost throughout India and found almost throughout the old world.

Habits and Habitat: A common species and mainly visible in spring and monsoons. Found from April to October, mostly in grass fields and meadow around the lake. This species can be recorded from the Khajjiar meadow mainly during spring and summer seasons.

Genus: *Pseudosphingonotus* Shumakov

1963. *Pseudosphingonotus* Shumakov, *Trud. Vsesoyuzu ent. obshch.* 49: 3-248.

6. *Pseudosphingonotus savignyi* (Saussure, 1884)

1884. *Sphingonotus savignyi* Saussure, *Mem. Soc. Geneve,* 28 (9): 198, 208; 1888. 30 (1): 78

1985. *Pseudosphingonotus savignyi* Bhowmik, *Rec. Zool. Surv. India.,* Occ. Paper No. 78: 39

Size: Length: 20-33 mm.

Diagnostic Characters: Body slender with grey or reddish colour. Colour of head and under surface varied with white; head prominent, vertex convex, antennae ringed with whitish colour. Pronotum constricted in front, obtusely rounded off behind. Tegmina with the basal third opaque, light reddish brown, very thickly reticulated; the rest subhyaline, generally with two slightly-indicated transverse light brown bands, and scattered brown spots beyond. Wings greenish hyaline, with a curved dark brown band, nervures of the wings colourless, except some of the longitudinal nervures. Hind femora pale, with three blackish bands above, the last band extending more or less on the sides.

Distribution: Found throughout the Western Himalaya. Also present in Pakistan, Central and Western Asia, North Africa.

Habits and Habitat: A very species common and mainly collected in spring and monsoons from the grassy fields and ground around the lake from May to October. During monsoons a good number of this species can be seen in Khajjiar pasture.

Subfamily: Hemiacridinae

Genus: *Spathosternum* Karsch

1877. *Spathosternum* Karsch, *Sitz. Akad. Wiss. Wien, Math.Nat. Cl.* lxxvi (1): 44

Tegmina with a patch of densely placed transverse veinlets between radial and median areas; apex of male cercus spine like.

7. *Spathosternum pr. prasiniferum* (Walker, 1871)

1871. *Heteracris prasinifera* Walker, *Cat. Derm. Salt. Brit. Mus.,* 5 (Suppl.): 65

1914. *Spathosternum prasiniferum* Kirby, *Fauna British India, Ortho.*: 208

Size: Length: 13-20 mm

Diagnostic Characters: It has broad blackish or dark green stripe running the lower part of the eyes and below the lateral carinae of the pronotum. Central area of tegmina with a longitudinal black streak, generally almost obsolete in the male and well marked in female, but very variable, sometimes being entire; dark brown spots on the elytra; spatulate type of prosternal tubercles; wing well developed.

Distribution: Found almost throughout India, found mostly near cultivated fields, grassland and on the periphery of forested areas. Elsewhere found in Bangladesh, Sri Lanka, Myanmar, Thailand, Indo-China, China and Vietnam.

Habits and Habitat: This species does not show any seasonal variation and is very common in status. Visible throughout the year where except months of extreme cold. Also a major pest of crops like wheat, paddy, beans, sugarcane, weeds, grasses and some garden plants. Present in Khajjiar meadow mainly during months of monsoon.

Subfamily: Oxyinae

Genus: *Oxya* Serville

1831. *Oxya* Serville, *Ann. Sci. Nat.* 22: 264, 286

External apical spine of hind tibiae absent; in male 10th abdominal tergite without furcula; male epiphallus without ancorae.

8. *Oxya fuscovittata* (Marschall, 1836)

1836. *Gryllus fuscovittatus* Marschall, *Ann. Wien Mus. Vienna*, 1 (2): 211

1971. *Oxya fuscovittata* Hollis, *Bull. Br. Mus. Nat. Hist.* (Ent.) 26 (7): 289

Size: Length: 18-21 mm.

Diagnostic Characters: Integument finely pitted and shiny. Circus of male compressed and weakly bifurcate at apex. Ventral surface of subgenital plate of female almost completely flat or weakly concave.

Distribution: Found in Andhra Pradesh, Arunachal Pradesh, Assam, Bihar, Goa, Himachal Pradesh, Jammu and Kashmir, Madhya Pradesh, Maharashtra, Orissa, Rajasthan, Uttar Pradesh and West Bengal. Elsewhere found in Afghanistan, Bangladesh, Pakistan and Russia.

Habits and Habitat: A very common species that does not show any seasonal variations. Found throughout the year everywhere. A pest of crops like paddy, potatos, sugarcane, green grass and pea. Found in good numbers from Khajjiar meadow during summer and monsoons.

Subfamily: Coptacrydinae

Genus: *Eucoptacra* Bolivar

1902. *Eucoptacra* Bolivar, *Ann. Soc. Ent. France*, lxx: 623, 625

Prosternal tubercle is stout and slightly acute.

9. *Eucoptacra saturata* (Walker, 1870)

1870. *Eucoptacra saturatum* Walker, *Cat. Derm. Salt. Brit. Mus.*, 4: 628

1921. *Eucoptacra saturata* Uvarov, *Ann. Mag. Nat. Hist.*, 9 (7): 503

Size: Length: Male 15 mm, Female 21-22 mm.

Diagnostic Characters: Overall colour ferruginous. Prosternal tubercle stout and slightly acute. Hind femora black beneath, as long as the abdomen; hind tibiae red; tegmina with numerous small marks; hind wings hyaline or grey tinged with black.

Distribution: Widely distributed species, found throughout India.

Habits and Habitat: A rare species, mostly collected in springs and late monsoons from June to October preferably forest and grass fields. Can be recorded from the meadow during late spring season.

Subfamily: Eyprepocnemidinae

Key to the genera

1. Posterior femur moderately long, not inflated basally; prosternal tubercle with rounded apex ....*Eyprepocnemis*
- Posterior femur long, inflated basally; prosternal process almost spathulate, with slightly inflated apex ....*Tylotropidius*

Genus: *Eyprepocnemis* Fieber

1853. *Eyprepocnemis* Fieber, *Lotos* 3: 98

10. *Eyprepocnemis rosea* Uvarov, 1942

1942. *Eyprepocnemis rosea* Uvarov, *Ann. Mag. Nat. Hist.* (11) 9 (56): 597

Size: Length: 19-27 mm.

Diagnostic Characters: Pronotum relatively short, obtusely tectiform above, median carina well developed. All sulci distinct, typical sulcus place behind the middle; prosternal tubercle tongue shaped. Last tergite with a broad and shallow median excision, flanked by weak projections. Supra-anal plate elongate; circus longer than the plate; subgenital plate short, obtusely conical; tegmina reaching the apex of abdomen plate in male, and to the middle of supra-anal plate in female; wings light rose basally.

Distribution: Found in Assam, Himachal Pradesh, Meghalaya and Uttar Pradesh.

Habits and Habitat: A very common species, mostly collected in springs and monsoons in the months of May to September. Generally present Khajjiar area in late spring and monsoons.

Genus: *Tylotropidius* Stal

1873. *Tylotropidius* Stal, *Recen. Ortho. Revue Crit. Ortho.dec. Linn.,* 1: 154.

11. *Tylotropidius varicornis* (Walker, 1870)

1870. *Heteracris varicornis* Walker, *Cat. Derm. Salt. Br. Mus.,* 4: 667

1914. *Tylotropidius varicornis* Kirby, *Fauna Brit. India.* Orth.: 265

Size: Length: 29-40 mm.

Diagnostic Characters: Pronotum brown with lateral carinae of pale colour; prosternal tubercle compressed, bituberclate at apex; hind femora thickened at the base, very slender towards the tip; hind tibiae and tarsi dull blue; supra anal plate of male elongate-triangular and sulcate; cerci straight, slightly compressed and acuminate.

Distribution: From India reported in the states of Andhra Pradesh, Goa, Himachal Pradesh, Maharashtra, Meghalaya, Orissa, Rajasthan, Tamil Nadu and West Bengal. Elsewhere occurs in Myanmar and Sri Lanka.

Habits and Habitat: This species does not show any seasonal variation and a very common in the suitable habitat types. Can be observed throughout the year. Mostly recorded from edges of Khajjiar meadow during late spring season.

Subfamily: Catantopinae

Key to the genera

1. Prosternal tubercle cylindrical and rounded at apex; pronotum subcylindrical..*Catantops*
- Prosternal tubercle neither cylindrical nor with rounded apex; pronotum not subcylindrical ....*Xenocatantops*

Genus: *Diabolocatantops* Jago

1984. *Diabolocatantops* Jago, *Trans. Amer. ent. Soc.*, 110: 370

12. *Diabolocatantops innotabilis* (Walker, 1870)

1870. *Acridium innotabilis* Walker, *Cat. Derm. Salt. Br. Mus.*, 4: 629

1953. *Catantops pinguis innotabilis* Dirsh & Uvarov, *Tijdschr. Ent.*, 96 (3): 2333

Size: Length: 27-34 mm.

Diagnostic Characters: Overall colour reddish-brown. Frontal ridge finely punctured, and slightly expanded between the antennae. Eyes approximating; antennae filiform. Pronotum closely punctured; carina slight continuous with well marked sulci. Tegmina extending beyond the abdomen; wings dull hyaline or greenish towards the tip. Hind femora stout, with two transverse black spots above; hind tibiae and tarsi red. Cerci of male slightly expanded at the tip.

Distribution: Found in Andhra Pradesh, Arunachal Pradesh, Assam, Bihar, Delhi, Goa, Laccadive Islands, Himachal Pradesh, Jammu and Kashmir, Kerala, Madhya Pradesh, Meghalaya, Orissa, Sikkim, Tamil Nadu, Uttar Pradesh and West Bengal. Outside India found in Afghanistan, Bangladesh, Borneo, Korea, Malaya, Maldive Islands, Sri Lanka, Myanmar, China, Cambodia, and Japan.

Habits and Habitat: An uncommon species and visible during the spring and monsoons from July to mid October in the forest fields.

Genus: *Xenocatantops* Dirsh and Uvarov

1953. *Xenocatantops* Dirsh & Uvarov,

13. *Xenocatantops karnyi* (Kirby, 1910)

(Plate 7 A)

1910. *Catantops karnyi* Kirby, *Syn. Cat. Orth.*, 3: 483

1982. *Xenocatantops karnyi* Jago, *Trans. Am. Ent. Soc.*, 108 (3): 455

Size: Length: 15-21 mm.

Diagnostic Characters: Apical bifurcation of circus, bluntly pointed and subequal; epiphallus distinctive with lophi almost square in plane view; male circus clearly birfurcate apically. Pronotum constricted in the middle; prosternal tubercle conical.

Distribution: Found in Andhra Pradesh, Delhi, Himachal Pradesh, Tamil Nadu and Uttar Pradesh.

Habits and Habitat: A very common species and collected in springs and monsoons. Found from March to September in the forest fields around the lake. A good population is present in all habitat types of Khajjiar meadow.

Subfamily: Cyrtacanthacrydinae

1. Prosternal tubercle very long and recurved 2
- Prosternal tubercle short and straight 3
2. Cerci of male rather short, compressed and laminated; subgenital lamina is triangularly emarginated ....*Cyrtacanthacris*
- Cerci of male long, not compressed, subgenital lamina is not triangularly emarginated...................................*Chondracris*
3. Tegmina with straight venation in apical part; hind tibiae with 8 spines on outer dorsal margin; prosternal tubercle almost cylindrical...*Patanga*

Genus: *Chondracris* Uvarov

1923. *Chondracris* Uvarov, *Bull. Ent. Res.*, 14: 38

14. *Chondracris rosea* (De Geer, 1773)

1773. *Acrydium roseum* De Geer, *Mem. pour Servir. des Insectes*, Stock. 3: 488

1923. *Chondracris rosea* Uvarov, *Bull. Ent. Res.* 14: 39

Size: Length: 45-80 mm.

Diagnostic Characters: Insect has large sized body, pale-green in colour. Antennae yellow. Scutellum of vertex hardly depressed; frontal ridge punctured above; pronotum tectiform; tegmina green; rose coloured basal area of hind wing; hind femora green, or with sides yellower; hind tibiae and tarsi purplish red.

Distribution: Found in Arunachal Pradesh, Assam, Bihar, Madhya Pradesh, Manipur, Tripura, Uttar Pradesh, West Bengal, Sikkim, Kerala and Tamil Nadu in India. Elsewhere present in Bhutan, Bangladesh, Thailand, Indonesia, Korea, Japan, Vietnam, China, Java, Tainan, Philippines, Manchuria and Hainan Islands.

Habits and Habitat: A rare species and mostly collected in springs and monsoons from April to September preferably in the forest areas. Also considered as a pest of rice and beans. Visible mostly during rains from the Khajjiar area.

Genus: *Cyrtacanthacris* Walker

1870. *Cyrtacanthacris* Walker, *Cat. Derm. Salt. Brit. Mus.*, 3: 550

15. *Cyrtacanthacris tatarica* (Linnaeus, 1758)

1758. *Gryllus locusta tatarica* Linnaeus, *Syst. Nat.* ed. x, i: 432

1923. *Cyrtacanthacris tatarica* Uvarov, *Bull. Ent. Res.*, 14: 39

Size: Length: 40-62 mm.

Diagnostic Characteres: Fastigium of vertex angular, in middle depressed; antennae filiform, as long as head and pronotum together; pronotum moderately tectiform, slightly constricted, integument fairly rugose or smooth, posterior margin angular; median carina of pronotum a little raised and crossed by three transverse sulci; posterior tubercle large, strongly curved backwards; elytra extend beyond the apex of posterior femora; super-anal plate of male elongate and acutely conical; valves of ovipositor moderately robust.

Distribution: In India reported from Andhra Pradesh, Bihar, Rajasthan, Jammu and Kashmir, Himachal Pradesh, Haryana, Uttar Pradesh, Orissa, Karnataka, Maharashtra, Meghalaya, West Bengal, Arunachal Pradesh, Manipur and Tripura. Also reported from Africa, Sri Lanka, Bangladesh, Thailand, Hainan, Philippines, Sumatra, Madagascar and Seychelles.

Habits and Habitat: A rare species and observed in May to September from the lake area. And is a minor pest of paddy, maize and vegetables like lady's finger, chillie etc. Species can be recorded from the Khajjiar meadow mainly during monsoon months.

Genus: *Patanga* Uvarov

1923. *Patanga* Uvarov, *Ann. Mag. Nat. Hist.* (9) 12: 364

16. *Patanga succincta* (Johansson, 1763)

(Plate 7 B)

1763. *Gryllus Locusta succinctus* Johansson, *Amoen. Acad.*, 4: 398

1923. *Patanga succincta* Uvarov, *Ann. Mag. Nat. Hist.* (9) 12: 364

Size: Length: 58-61 mm.

Diagnostic Characters: Tegmina with straight venation in apical part, transverse veins forming almost right angle with principal veins; wing base tinged with pink colour; male subgenital plate long, curved upwards and its apex pointed.

Distribution: Found in Arunachal Pradesh, Assam, Bihar, Delhi, Goa, Himachal Pradesh, Jammu and Kashmir, Laccadive Island, Maharashtra, Meghalaya, Orissa, Rajasthan, Tamil Nadu, Uttar Pradesh and West Bengal. Elsewhere present in Saudi Arabia, Borneo, China, Hainan, Japan, Java, Malaya, Myanmar, Philippines, Sri Lanka, Sumatra, Taiwan and Tonkin.

Habits and Habitat: A common species and mostly collected in springs and monsoons. Species is a major pest of citrus plants. Visible from May to October. Present in good numbers from the grassy areas of the Khajjiar lake.

Family: Pyrgomorphidae

Key to the genera

1. Posterior lobe of pronotum convex, raised above the level of anterior lobe, with strong rugae; abdomen with dorsal callosities..*Aularches*

- Posterior lobe of pronotum on the level of anterior lobe; abdomen without callosities; outer apical spine of posterior tibiae distinct..*Poekilocercus*

Genus: *Aularches* Stal

1873. *Aularches* Stal, *Oefv. Vet. Akad. Forh.* 30 (4): 51

17. *Aularches punctatus* (Drury, 1773)

(Plate 7 C)

1773. *Gryllus (Locusta) punctatus* Drury, *Ill. Exot. Ent.* ii, pl. xli, fig.4

Size: Length: 42-60 mm.

Diagnostic Characters: This species has almost entirely, shining black colour above; a broad yellow band running across the face below the antennae and across the sides of pronotum; abdomen banded with yellow or reddish; legs black; hind femora sometimes yellowish; tegmina light brown, thickly reticulated with yellow; wings purplish brown and subhyaline.

Distribution: Found in Jammu & Kashmir, Uttarakhand and Uttar Pradesh in India and outside present in Tibet, Java and Nepal.

Habits and Habitat: Mainly collected in spring and monsoons and common in status. Present from May to October preferably in the shrubs. These are generally observed during late spring season from Khajjiar meadow.

Genus: *Poekilocerus* Serville

1831. *Poekilocerus* Serville, *Ann. Sci. Nat.* xxii: 275; 1839. *id., Ins. Orth.*: 595

18. *Poekilocerus pictus* (Fabricius, 1775)

(Plate 7 D)

1775. *Gryllus pictus* Fabricius, *Syst. Ent.*: 289

1914. *Poekilocerus pictus* Ambar, *Sind. Univ. Sci. Res. Journal, Karachi*: 1

Size: Length: 43-61 mm.

Diagnostic Characters: Colour blue-black or green with yellow markings and red wings. Antennae blue-black, ringed with yellow beyond basal third of the length. Head and pronotum with a slight median carina; head with a broad yellow band within each eye, running back on the prtonotum to the middle sulcus, behind which two broad subinterrupted transverse yellow bands; broad yellow diverging bands on the sides of frontal ridge; a broad yellow band present below each eye. Pronotum impress-punctate, rounded behind. Tegmina green or olive with the longitudinal and transverse nervures yellow and apex reddish; wings brick-red, subhyaline towards the tip. Abdomen yellow, with transverse blue-black bands. Legs yellow; the four tibiae blotched with blue-black.

Distribution: Found in Andhra Pradesh, Karnataka, Orissa, Maharashtra, Rajasthan and West Bengal. Also present in neighbouring country Pakistan.

Habits and Habitat: A common species that does not show any seasonal variation. Found throughout the year except December, January and February, mostly in the agriculture field and forest. A serious pest of ber, banana, aaks, calotropis spp., fig, grapevine, guava, melons, papaya, peach and various ornamental plants and vegetables. Present in good numbers from the Khajjiar meadow.

Infraorder: Tetrigidea

Superfamily: Tetrigoidea

Family: Tetrigidae

Pronotum almost covers the whole abdomen and conceals the hind wings; tegmina reduced to small lateral scales; hind wings well developed; anterior and middle tarsi two segmented, posterior three segmented; empodium absent between claws; epiphallus present.

Key to the Subfamilies

1. Posterior angle of lateral lobes of pronotum acutely produced outwards and generally spined; first segment of posterior tarsi usually longer than the third..............................Scelimeninae

- Posterior angle of lateral lobes of pronotum rarely acutely spined and turned downwards; first segment of posterior tarsi not longer than the third...2

2. Posterior angles of lateral lobes of pronotum little produced outwards obliquely truncated behind, very rarely acutely spinose; first and third posterior tarsal segments nearly equal in length ....Metrodorinae

3. Posterior angles of lateral lobes of pronotum turned downwards, more or less rounded; first and third posterior tarsal segments equal in length, first rather longer than the third ....Tetriginiae

Subfamily: Scelimeninae

Genus: *Eucriotettix* Hebard

1929. *Eucriotettix* Hebard, *Rev. Suisse Zool.*, 36: 573

Generally medium or large sized; posterior angle of lateral lobes of pronotum acute and produced into a spine at the anterior margin which is more or less transverse or obliquely directed forwards.

19. *Eucriotettix grandis* (Hancock, 1915)

1915. *Criotettix grandis* Hancock, *Mem. Dip. Agric. India Ent. Ser.*, 4: 134

1938. *Eucriotettix grandis* Gunther, *Stettin. Ent. Ztg.*, 99: 182

Size: Length: 17-20 mm.

Diagnostic Characters: Head a little elevated; vertex narrower than an eye, a little narrowed in front; frontal costa moderately divergent behind the paired ocelli; pronotum angularly excavate in the middle, otherwise truncate anteriorly; dorsum granulose; wings extended up to the apex of pronotum; first segment of posterior tarsi longer than the third.

Distribution: Found in Arunachal Pradesh, Assam, Meghalaya, Sikkim and West Bengal. Elsewhere present in Nepal and Myanmar.

Habits and Habitat: A common species and found from March to October. Observed in good numbers during rainy season in Khajjiar area.

Subfamily: Metrodorinae

Genus: *Teredorus* Hancock

1906. *Teredorus* Hancock, *Ent. News*, 18: 52

20. *Teredorus frontalis* Hancock, 1915

1915. *Teredorus frontalis* Hancock, *Rec. Ind. Mus.*, (Ent.), 11: 110

Size: Length: 11-15 mm

Diagnostic Characters: Head a little exerted above the surface of pronotum. Vertex strongly narrowed in front; frontal costa bifurcate behind the paired ocelli; antennae inserted below the eyes; pronotum extending beyond the hind femoral apices; posterior angle of lateral lobes of pronotum turned downwards; wings extending upto the pronotal apex.

Distribution: Found in Arunachal Pradesh, Assam, Himachal Pradesh, Manipur, Meghalaya, Sikkim and West Bengal.

Habits and Habitat: A common species and found in spring and monsoons from March to October. Recorded in Khajjiar area mainly during monsoon months.

Subfamily: Tetriginae

Key to the genera

1. Vertex and eyes obviously raised above the level of pronotum; antennae inserted between the lower border of eyes, antennae inserted distinctly below the inferior margin of eyes......*Ergatettix*
- Vertex and eyes very little or not at all raised above the level of pronotum; antennae scarcely placed below the middle of eyes.....2
2. Body generally smooth or little granulose; vertex equal to or a little narrower than an eye, carinated in front; median carina of pronotum percurrent, not at all interrupted ....*Hedotettix*

- Body generally granulose or tuberculose; vertex not expanded but narrowed in front; abbreviated carinae indistinctly present...*Coptotettix*

Genus: *Coptotettix* Bolivar

1887. *Coptotettix* Bolivar, *Ann. Soc. Ent. France*, 31: 287

21. *Coptotettix conspersus* Hancock, 1915

1915. *Coptotettix conspersus* Hancock, *Rec. Indian Mus.*, 11: 119

Size: Length: 11-16 mm.

Diagnostic Characters: Vertex not produced in front of eyes; fossulae present; frontal costa wide between antennae; pronotum and wings extend beyond the hind femoral apices; dorsum convex with abbreviated curved carinae; median carina gently undulate behind; third pulvilli either equal to or longer than the previous two.

Distribution: Found in Northwest, Central and Northeast India. Also present in Sri Lanka.

Habits and Habitat: Found from May to October preferably in the agriculture and grasses field. A common species in status and mostly found in spring and monsoons.

Genus: *Ergatettix* Kirby

1914. *Ergatettix* Kirby, *Fauna Brit. Ind.*, Ortho. 1: 69

22. *Ergatettix dorsiferus* (Walker, 1871)

(Plate 7 E)

1871. *Tettix dorsiferus* Walker, *Cat. Derm. Salt. Brit. Mus.*, 5: 825

1929. *Ergatettix dorsiferus* Hebard, *Revue Suisse Zool.*, 36: 588

Size: Length: 9-13 mm.

Diagnostic Characters: A medium sized species. Head distinctly exserted above the surface of pronotum; vertex narrower than an eye; antennae situated below the inferior margin of eyes; pronotum and wings extend beyond the apex of hind femora; dorsum wide between shoulders; median carina depressed in front, undulate behind the shoulders; posterior femora elongate; first and second pulvilli small and spinose third equal to or longer than second.

Distribution: Found almost throughout India. Also recorded from Bangladesh, Burma, Central Asia, South China, Sri Lanka, Sumatra and Taiwan.

Habits and Habitat: Found throughout the year except winters, everywhere, mostly in the agriculture and grass field.

A very common species in the lake area.

23. *Ergatettix guentheri* Steinmann, 1970

1970. *Ergatettix guntheri* Steinmann, *Acta zool. Hung.*, 16: 234

Size: Length: 9-13 mm.

Diagnostic Characters: Median carina strongly undulate behind shoulders; lateral carinae also wavy with nodules especially towards the apex of pronatal process; posterior femora elongate, crassate, external surface with distinct projecting tubercles; first and second pulvilli small and spinose, third equal to or longer than the second.

Distribution: Found in Bihar, Madhya Pradesh, Maharashtra, Orissa, Uttar Pradesh, West Bengal and Northeast Indian States. Elsewhere present in Bangladesh and Sri Lanka.

Habits and Habitat: Found throughout the year except winters months. A very common species, visible everywhere, mostly in the agriculture field.

Genus: *Hedotettix* Bolivar

1887. *Hedotettix* Bolivar, *Ann. Soc. Ent. Belgique,* 31: 283

24. *Hedotettix costatus* Hancock, 1912

(Plate 7 F)

1912. *Hedotettix costatus* Hancock, *Mem. Dep. Agric. India, Ent. Ser.,* 4: 147-148

Size: Length: 11-13 mm.

Diagnostic Characters: A medium sized species. Head not exserted above the pronotum. Vertex equal to or a little narrower than an eye, expanded, angularly produced in front, carinate in middle; frontal costa gently furcated. Pronotum extend beyond the hind femoral apices; dorsum finely granulose. Wings extend beyond the pronotal apex.

Distribution: Outside India present in Bangladesh and Nepal. Found in Northwest, Central and Northeast India.

Habits and Habitat: Found from March to October mostly in the grass field. A common species in Khajjiar area.

Suborder: Ensifera

Superfamily: Grylloidea

Key to the families

1. Posterior tibiae without denticles between the spines......Gryllidae
2. Posterior tibiae non-serrulated, armed with three spines on each margin; very small delicate insects.......................Trigonidiidae

Family: Gryllidae

SubFamily: Gryllinae

Spines of the posterior tibiae immovable, without hairs.

Key to the genera

1. Elytra of male with well developed mirror 2
- Elytra of male with mirror indistinct or small, situated towards apex 3
2. Face of male strongly flattened and forehead more or less strongly prolonged; frontal rostrum of male projecting but simply angular ....*Loxoblemmus*
- Male with flattened or somewhat concave face, but forehead without any prolongation; frontal rostrum convex ....*Velarifictorus*
3. Anterior projection of the ectoparamere approximately equal...*Gryllus*
- Anterior projection of the ectoparamere much longer than the external anterior projection, Mesal lobes with their apices between the apices of the ectoparameres ....*Teleogryllus*
4. Endoparameres elongate, J-shaped ....*Acheta*

Genus: *Acheta* Fabricius

1775. *Acheta* Fabricius *Systema Entomologiae.* Leipzig, korie pp: 832

25. *Acheta domesticus* Linnaeus, 1758

1758. *Gryllus acheta domesticus* Linnaeus, *Syst. Nat.*, (ed. x), 1: 428

1967. *Acheta domesticus* Chopard, *Orthopterorum Catalogus*, pt. 10: 62

Size: Length: 16-20 mm.

Diagnostic Characters: Reddish yellow, medium sized species. Body somewhat depressed and pubescent; head with a white transverse band.

Distribution: Found in West Bengal, Jammu and Kashmir, Kerala, Maharashtra and Uttar Pradesh. Elsewhere found in Pakistan.

Habits and Habitat: A very common species, mostly found in spring and monsoon seasons. In Khajjiar area, can be observed almost throughout the year except winters, everywhere, mostly in the agriculture field.

Genus: *Gryllus* Linnaeus

1758. *Gryllus* Linnaeus, *Syst. Nat.* ed. x, i: 425

26. *Gryllus bimaculatus* De Geer, 1773

1773. *Gryllus bimaculatus* De Geer, *Mem. Ins.*, 3: 521

Size: Length: 21-30 mm.

Diagnostic Characters: A large sized species with completely glabrous body; general colouration entirely black with two yellow spots at the base of each elytron.

Distribution: Found in West Bengal, Andaman Islands, Bihar, Jammu and

Kashmir, Karnataka, Madhya Pradesh, Maharashtra, Tamil Nadu and Uttar Pradesh. Elsewhere found in Africa, Myanmar, Pakistan, Singapore and Sri Lanka.

Habits and Habitat: A very common species without seasonal variations. One of the major pests of potato and paddy. Found throughout the year, mostly in the agriculture field and grass land in Khajjiar area.

Genus: *Loxoblemmus* Saussure

1877. *Loxoblemmus* Saussure, *Mem. Soc. Geneve*, 25: 249

27. *Loxoblemmus equestris* Saussure, 1877

(Plate 7 G)

1877. *Loxoblemmus equestris* Saussure, *Mem. Soc. Geneve*, 25: 252

Size: Length: 11-13 mm.

Diagnostic Characters: Head with 6 yellow lines on occiput and a transverse band at the top of frontal rostrum; face wide, flattened, strongly oblique; antennae without any projection; elytra on dorsal field has a reticulation formed of irregular lengthened aerolae.

Distribution: Found in West Bengal, Assam, Bihar, Jammu and Kashmir, Karnataka and Punjab. Elsewhere in China, Myanmar, Japan and Indo-Malaya.

Habits and Habitat: Found from March to October preferably in the agriculture field. A common species, mostly seen in spring and monsoons from Khajjiar area.

Genus: *Teleogryllus* Chopard

1961. *Teleogryllus* Chopard, *Eos. Madr.*, 37: 277

28. *Teleogryllus occipitalis* (Serville, 1838)

(Plate 7 H)

1838. *Gryllus occipitalis* Serville, *Handb. Ent.* : 339

1980. *Teleogryllus occipitalis* Townsend, *J. Nat. Hist.*, 14; 154

Size: Length: 20 mm.

Diagnostic Characters: Head with yellow band along the internal margin of eyes; mirror broader than long, divided in the middle by curved vein, 4 oblique veins; genitalia with large superior part, sinuated sides, and feebly acute at apex.

Distribution: Found in Andaman Islands, Arunachal Pradesh, Assam, Bihar, Himachal Pradesh, Karnataka, Madhya Pradesh, Manipur, Meghalaya, Orissa, Sikkim, Tamil Nadu, Uttar Pradesh and West Bengal. Also present in countries like Bhutan, Japan, Laccadive, Maldive, Sri Lanka, Philippines and Tibet.

Habits and Habitat: Found throughout the year except months of November to January. A very common species without any seasonal variations.

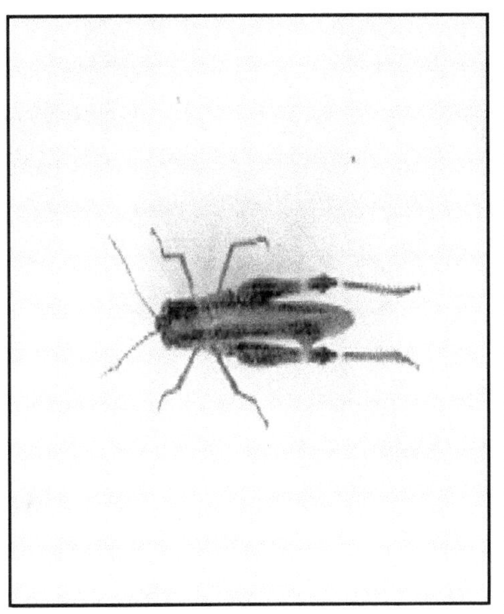

A ***Ceracris n. nigricornis***

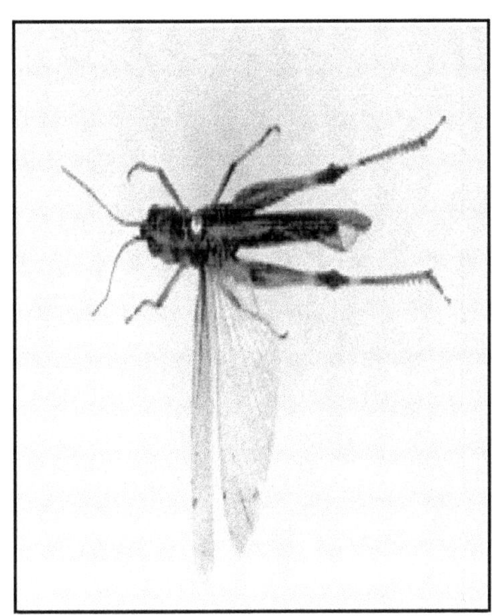

B ***Ceracris n. nigricornis***

C ***Dnopherula (Aulac.) decisus***

D ***Spathosternum pr. Prasiniferum***

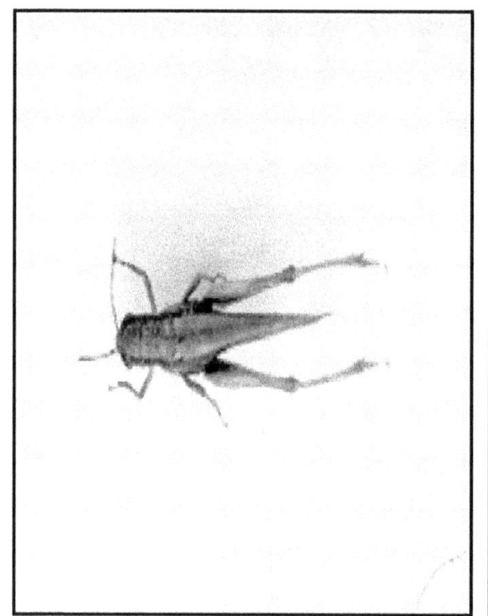

E *Oxya fuscovittata*

F *Eyprepocnemis rosea*

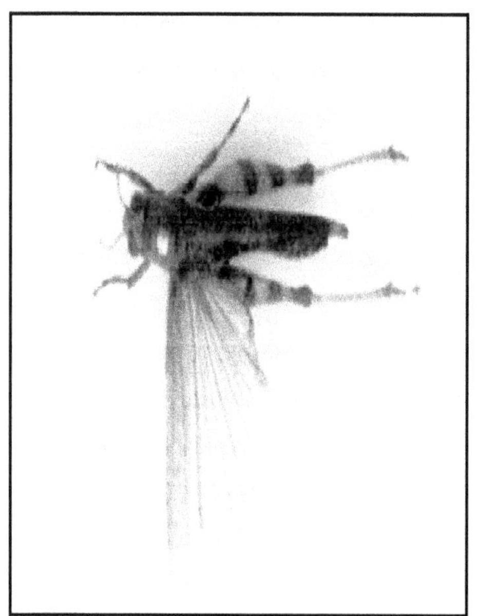

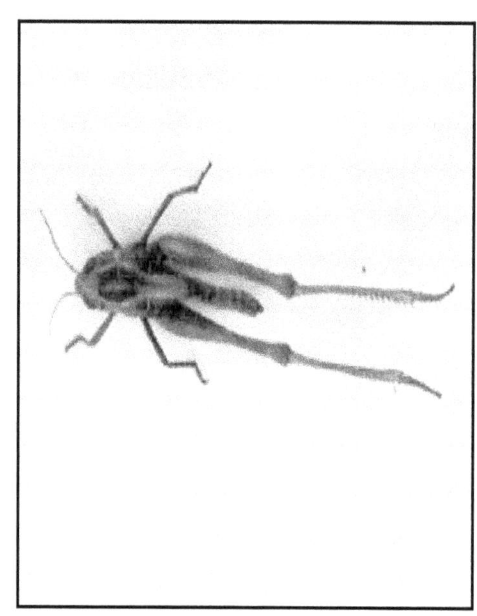

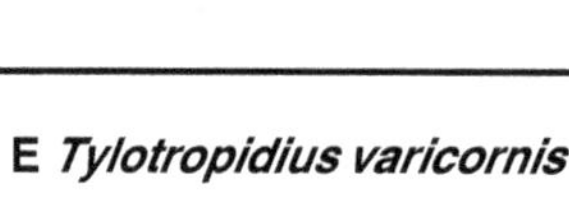

E *Tylotropidius varicornis*

F *Diabolocatantops innotabilis*

Plate 5.7: Some orthopteran recorded in Khajjiar lake area

A *Xenocatantops karnyi*

B *Patanga succinct*

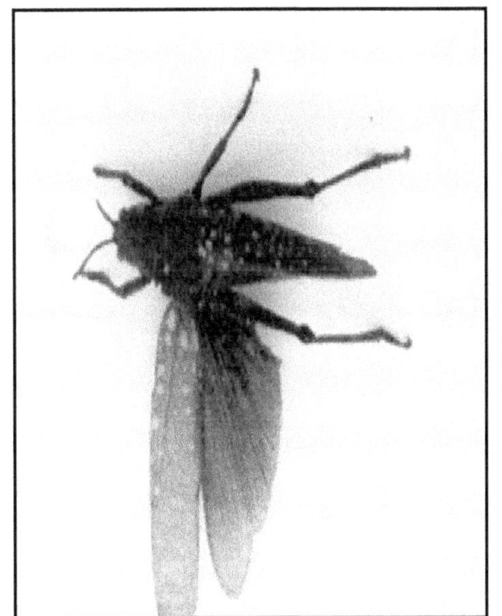

C *Aularches punctatus*

D *Piokilocerus pictus*

E *Ergatettix dorsiferus*

F *Hedotettix costatus*

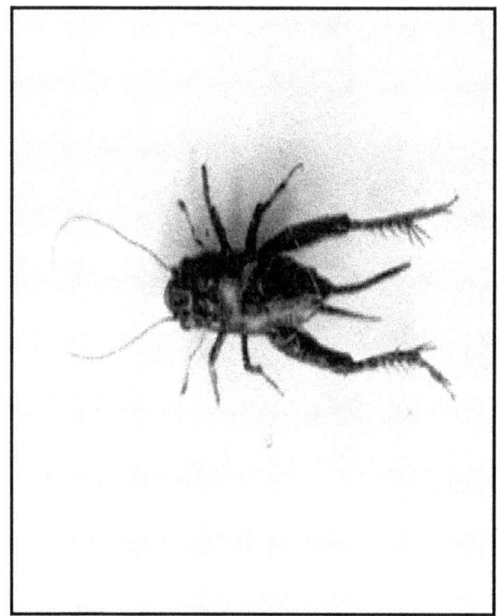

G *Loxoblemmus equestris*

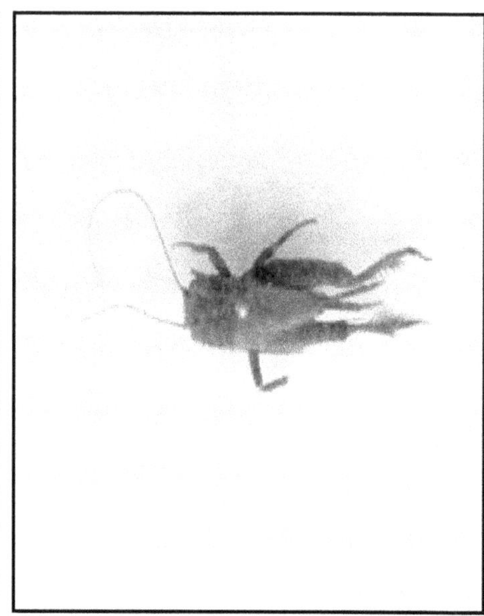

H *Teleogryllus occipitalis*

Plate 5.8: Some orthopterans recorded in Khajjiar lake area

Superfamily: Tettigonioidea

Family: Conocephalidae

Fore and middle femora spined below; male subgenital plate bifurcated at apex.

Genus: *Conocephalus* Thunberg

Small species with a short rounded rostrum; fore and middle femora unarmed; fastigium of vertex broad.

29. *Conocephalus maculatus* (Le Guillou, 1841)

1841. *Xiphidion maculatum* Le Guillou, *Revue Suisse. Zool.*, 4: 294

1980. *Conocephalus maculatus* Pitkin, *Bull. Br. Mus. Nat. Hist.* (Ent.) 41 (5): 344

Size: Length: 12-15 mm.

Diagnostic Characters: Fastigium of vertex moderately wide, pronotum bispinose; tegmina pigmented with comparatively large dark spots; hind femora inarmed ventrally; ovipositor very shot, relatively straight.

Distribution: Found throughout India. Elsewhere present in Australia, Cameron, Indonesia, Java, Western Malayasia, Philippines and Sierra Leone.

Habits and Habitat: Prefers grass fields in suitable terrains. Found from March to November in Khajjiar preferably on the grass field and very common in status.

ost covers the whole abdomen and conceals the hind wings; tegmina reduced to small lateral scales; hind wings well develo

C). Order: Hemiptera

Hemelytron type of the wing present in the order with basal portion of the front wing thickened and leathery, and the apical potion membranous. The hind wings entirely membranous and slightly shorter than the fore wings. The wings at rest, kept flat over the abdomen, with the membranous tip of the front wings overlapping. Mouth parts, of the piercing sucking type and in the form of slender, usually segmented beak that arises from the front part of the head and extends back along the ventral side of the body, sometimes as far as the base of hind wings. Segmented portion of the beak called labium serves as a sheath for the four piercing stylets.

In khajjiar area 5 species of hemipterans belonging to 4 genera and 2 families are present. Analyses of data reveals that family Pentatomidae with 3 species over numbered the family Largidae (2 species) in terms of number of species in the area.

Family: Pentatomidae

Members of this family easily recognised by their shield like shape and five segmented antennae. Most of these bugs brightly coloured and conspicuously marked therefore also called shield bugs. The tarsi 3-segmented. The forewings with the basal half thickened while the apex membranous, like the hindwings. Most of these insects eject a foul smelling glandular substance.

Genus: *Dalpada* (Amy and Serc.)

1843. *Dalpada*, Amy and Serc. *Hem.* p. 105

Body oval in shape, head slightly convex and lateral lobes more or less distinctly sinuate at their apices. Bucculae angulated anteriorly and reach the base of the head. Antennae five-jointed and slender. Eyes globose and prominent. Rostrum extends beyond the posterior coxa. Abdomen moderately sulcated on basal area. Tibiae furrowed, anterior tibia divided.

1. *Dalpada nigricollis*, Westwood

(Plate 8 A)

1837. *Dalpada nigricollis*, Westwood (Halvs) *In Hope Cat. Hem.* i. p. 22

Size: Length 15-18mm; breadth 8- 9 mm.

Diagnostic Characters: Overall greyish-luteous and very thickly punctured with brassy black colour. Pronotum with a slender discal median line, four minute ochraceous spots in a transverse series in the front. Anterio-lateral margins of the pronotum serrate, posterior angles prominent and slightly recurved inwardly with two furrows. Underside of the body dark ochraceous with darkly punctuate lateral areas. Antennae fuscous and base of fourth and fifth joints ochraceous.

Distribution: Reported from Punjab, Jammu and Kashmir, Himachal Pradesh and Bengal. Elsewhere occurs in Nepal and west China.

Habits and Habitat: A common species found in grassy areas in suitable habitat types throughout their range of distribution. This is an uncommon species in the Khajjiar present in grassy margins of the lake.

Genus: *Erthesina* Spin

1837. *Erthesina*, Spin. *Ess.* p. 291

Anterior and posterior tibiae dilated. Head long and tapering at apex. Basal joint of the antennae not reaching the apex of the head. Basal joint of the rostrum extends beyond the bucculae. Veins of the membrane simple.

2. *Erthesina fulla* (Tlmnb, 1783)

(Plate 8 B)

1783. *Erthesina fulla*, (Tlmnb) (Cimex), *Nov. Ins. Spec.* ii, p. 42

Size: Length 20-25 mm; breadth: 11-12 mm

Diagnostic Characters: Head black and coarsely punctuate with a central longitudinal line, and chraceous inner margins of eyes. Antennae piceous and base of apical joint ochraceous. Pronotum and scutellum black and very coarsely punctate, with scattered small ochraceous callosities. Corium purplish brown with marginal area somewhat darker. Underside of body pale ochraceous, with coarse scattered black punctures on the lateral areas. Legs black, bases of femora, centres of tibiae, and bases of tarsi all ochraceous.

Distribution: In India found in Assam, Bengal, Kerala, Andaman Islands and Himachal Pradesh. Elsewhere, present in China and Japan.

Habits and Habitat: A pest of Pine and hard wood trees in Asia. Considered as agricultural pests because they can produce large populations; suck plant juices and damage crop production. Abundant in pine forests on the edges of the Khajjiar meadow, winged ones sometimes recorded from the meadow area as well.

Genus: *Prionolomia*, Stdl.

1873. *Priouolomia*, Stdl, *En. Hem.* iii, p. 37

Lacks the lobate process to the sub-apical upper surface of the femora-distinguishing feature from other genera of same the family. First joint of the antennae not long than the fourth. Posterior tibia are dilated and in the male denticulate on the inner side. Posterior femora in the male with a strong spine before apex. Abdomen unarmed.

3. *Prionolomia cardoni* Lethierry, 1891

(Plate 8 C)

1891. *Prionolomia cardoni*, Lethierry, *Bull Soc. Ent. Belg.* p. Xliii

Diagnostic Characters: Overall colouration very deep. Anterior area of pronotum granulate. Pronotum, scutellum and membrane deeper piceous-brown. Body beneath in female obscure castaneous, in male with the sternum piceous, the sterna segmental margins and the abdomen castaneous. Lateral expansion of the pronotum more rounded and enlarged, the dentations more acute, the granules on the femora smaller and more numerous. The angles on the inner margin of the tibiae nearer the base and more obtuse. Posterior femora in the male strongly incrassated, tuberculate on each side.

This species can be seen most of the times from the fencing area of the glade in Khajjiar.

Family: Largidae

Members of this family have stout and wide-bodied, with no ocelli, and four-segmented rostrum. They have contrasting coloured edges to their elytra. Generally ground-dwelling or scramble around in plants, bushes and trees. Feed on plant juices and seeds.

Genus: *Physopelta* Amy and Serv.

1843. *Physopelta*, Amy.and Serv. *Hem.* p. 271

Body oblong in shape; head somewhat large, equilateral and convex. First joint of antennae in both sexes longer than head, but shorter than head and pronotum together. Anterior convex area of the pronotum reaches the anterior margin, and the lateral pronotal margins less reflexed. Anterior femora strongly spined.

4. *Physopelta gutta* Burm., 1834

(Plate 8 D)

1834. *Physopelta gutta*, Burm. *Nov. Act. Acad. Leop.* xvi, p. 300

Size: Length: 15-17 mm

Diagnostic Characters: A medium sized bug, brown in general colouration with yellow or ochraeous and black patterns. Head small and yellow, vertex with a large brown spot. Head with a pair of well defined compound eyes and a pair of dark brown antennae. Base of apical joint of antennae ochraceous and base of first joint of antennae dull reddish. Pronotum, scutellum and basal of corium fuscous; a discal rounded spot and apical angles of corium and the membrane black. Coxae, trochanters and femora dull reddish beneath. Legs relatively long. Fore femora strongly widened and armed with two rows of alternatively large and small spines. Mid femora armed with two rows of small tubercles.

Distribution: Widely present in Australian, Oriental and Palaearctic regions. In India reported from Assam, Jammu & Kashmir, and Himachal Pradesh.

Habits and Habitat: Extracts juice from the seeds and also sucks sap from the stems. As a result, seeds fail to ripen and stem becomes stunted. Recorded in large numbers during June-July from the Lake area.

5.*Physopelta schlanbuschi* Fabricius, 1787

(Plate 8 E)

1787. *Physopelta schlanbuschi*, Fabricius *Mant. Ins.* ii, p. 299

Size: Length: 13-16 mm

Diagnostic Characters: A reddish-ochraceous coloured bug. Antennae large and pilose with apical joints greyish, first, second joints almost sub-equal in length. Two spots present on anterior lobe of pronotum. Two large transverse spots also present near anterior margin of posterior lobe. Scutellum with a rounded discal spot; a lateral series of long transverse linear spots present on sternal and abdominal incisures. Apex of rostrum, tibiae and tarsi black, base of first joint of antenna and apex of scutellum sanuineous. Posterior area of pronotum sparingly but very coarsely punctuated. Clavus somewhat coarsely and corium much more finely punctuate.

Distribution: Found in Myanmar, China and almost throughout India.

Habits and Habitat: Feeds on juice extracts of the seeds and on sap of stems. Generally this species is recorded in large numbers during June-July from the Lake area.

D). ORDER: HOMOPTERA

Insects of this order generally have four wings; the front wings have uniform texture throughout, either membranous or slightly thickened, and the hind wings membranous. The wings at rest usually held roof like over the body, with the inner margins overlapping slightly at the apex. Mouth parts similar to that of Hemiptera, sucking type with four pairs of piercing stylets. The beak arise from the back of the head, in some cases appearing to arise between the front coxae, in Hemiptera the beak arise from front of the head.

A single species of Homoptera (*Platylomia saturate*) belonging to family Cicadiae can be seen Khajjiar lake area. A very large population of this species is recorded mainly from the forest surrounding the meadow.

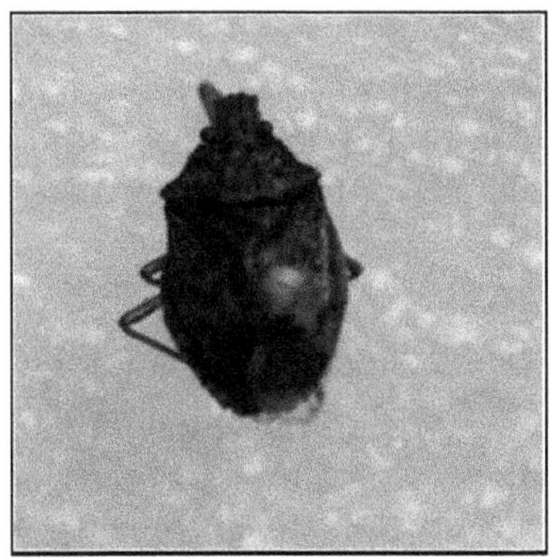

A *Dalpada nigricollis*

B *Erthesina fulla*

C *Prionolomia cardoni*

D *Physopelta gutta*

E *Physopelta schlanbuschi*

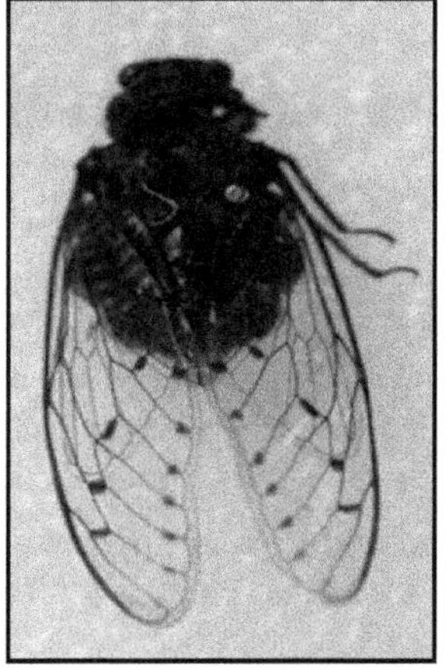

F *Platylomia saturata*

Plate 5.9: Some hemipterans recorded in Khajjiar lake area

Family: Cicadiae

Members of this family easily recognised by characteristic shape and large size. Cicadas have prominent eyes, set wide apart on the sides of the head. Antennae short, protruding in front of the eyes. Front wings membranous. Louder sound production-unique feature of the family.

Genus: *Platylomia* Stal

1870. *Platylomia*, Stal, *Ofv. Vet.-Ak. Fork.*, p. 708

Head little wider than base of mesonotum and almost as long as breadth between eyes. Pronotum at centre and almost as long as mesonotum. Abdomen longer than space between apex of head and base of cruciform elevation. Tympana completely covered. Tegmina and wings hyaline, the venation sometimes fuscously spotted.

1. *Platylomia saturata* (Walker, 1858)

(Plate 8 F)

1858. *Dundubia saturata* Walker, *List. Sp. Homopt. Ins. Col. Brit. Suppl.*: 6

Diagnostic Characters: Head and thorax dorsally olivaceous. Two central longitudinal fasciae on pronotum. Mesonotum with a central longitudinal fascia, connected with two large anterior fused spots. Abdomen shining pieous, lateral areas of segmental margins broadly ochraceously pilose. Head on underside, sternum and legs olivaceous. Opercula with an ochraceous tint. Underside of abdomen ochraceous. Tegmina and wings transparent, venation ferruginous.

Distribution: Distributed in India and Nepal. In India reported from Ranikhet, Sikkim, Assam, Naga hills, Himachal Pradesh and Jammu & Kashmir.

Habits and Habitat: Easily noticed due to production of very loud sound especially during breeding seasons which can be heard from distance in pine forests. Large population of this species can be observed from forest areas surrounding the Khajjiar meadow.

E) ORDER: COLEOPTERA

Structure of wing-most distinctive feature of this order. Most of the beetles have four wings, with the front pair thickened, leathery, or hard and bristle. Wings meet in a straight line down the middle of the back and covering the hind wings. Hindwings membranous, usually longer than the forewings, and when at rest, usually folded up under the forewings. Frontwings are called elytra and serve as protective sheaths. Hindwings ordinarily used for flight. Mouth parts chewing type and the mandibles well developed. Beetles undergo complete metamorphosis.

Khajjiar and surrounding area of Chamba district supports 15 species of beetles belonging to 15 genera and 7 families. These include some species of economic importance like *Carabus boysi, Dyticus punctalis, Mimela pectoralis, Onitis subopacus,* ***Melolontha furcicauda,*** *Anomala rufiventris, Coccinella septumpunctata and Mylabris pustulata.* Population of beetles in Khajjiar meadow increases during spring season and a multifold increase in their numbers is recorded during this season.

Family: Carabidae

Generally known as ground beetles. Members of this family show considerable variation in size, shape and colour but most of the species dark, shining and somewhat flattered with striated elytra. The elytra fused in some species, particularly large ones rendering the beetles unable to fly. Antennae arise more laterally on sides of head between eyes and base of mandibles. Clypeus not produced laterally beyond base of antennae.

Genus: *Calosoma* Weber.

1801. *Calosoma*, Weber, *Obs. Ent.* i, p. 20.

1. *Calosoma* sp.

(Plate 9 A)

Diagnostic Characters: Head wide and not contracted behind the eyes. Neck inflated; gula with a single seta on each side. Eyes prominent, distant from bucccal fissure. Labrum short but wider than clypeus, deeply emarginated in front. Mandibles long and powerful, slightly curved at the apex. Mentum emarginated, with a short but sharp tooth. Maxillae strong, densely fringed with hairs along the inner margins and over the whole of rounded apex. Antennae pubescent from joint 5. Joint 2 short, 3 very long, both compressed. Joints 1 and 4 sometimes with feeble edges. Prothorax short with rounded sides. Elytra wide and without basal border. Prosternum process bordered, produced over the mesosternum.

Habits and Habitat: Found in grassy areas with some busy vegetation. Good population of this species is present in grassy areas around Khajjiar lake.

Genus: *Carabus* Linnaeus

1758. *Carabus*, Linnaeus. *Syst. Nat. ed.* x, p. 418

Beautifully coloured insects, with metallic sheen. Head of different species variable in dimensions, sometimes inflated behind, sometimes with neck constriction. Eyes prominent and at a distance from the buccal fissure. Clypeus small, labrum well developed often wider than the clypeus. Mandibles powerful, their form and length varies according to species but generally curved and sharp at the apex; its inner margin clothed with hairs. Mentum with sharp tooth. Maxillae sharply hooked at the apex and fringed with hairs. Prothorax generally cordate, sometimes quadrate with bordered sides. Elytra without basal borders.

2. *Carabus boysi* Tatum, 1851

1851. *Carabus boysi*, Tatum. *Ann. Mag. Nat. Hist.* (2), viii, p. 51

Size: Length: 23-38 mm. Width: 85-135 mm.

Diagnostic Characters: Head wide and convex; colour of body dull black. Eyes prominent and finely wrinkled. Neck coriaceous having one elytra and one supraorbital seta on each side. Prothorax cordate and slightly convex. Apical border present, sides with thick and reflexed border, well rounded in front. Strongly sinuated at the fourth from the base; hind angles sharp and projecting backward and downwards. Elytron convex and elongate-ovate. Metasternum longer than wider. Venter more or less punctuating at sides. The segments plurisetone on each side of middle line. Front tarsi in males with four dilated joints. The mesotibia hardly produced at the apex.

Distribution: Found in Punjab, Himachal Pradesh (Chamba, Kullu and Spiti), Garhwall, Almora and some other parts of India.

Habits and Habitat: Found in grassy areas. This species is generally observed in grassy areas around Khajjiar meadow.

Family: Dytiscidae

Aquatic beetles which common in ponds and quiet streams. They have smooth body, somewhat oval in shape and very hard; dark brown, blackish or dark olive in colour with golden highlights in some species. They have short, but sharp mandibles. Hind legs flattened and fringed with long hairs to form paddles; can remain submerged under water for long periods because they carry air in their chambers under the elytra.

3. *Dytiscus punctalis* Linnaeus, 1758

(Plate 9 B)

1758. *Dytiscus punctalis* Linnaeus, *Syst. Nat.*, 10th ed., 475

Diagnostic Characters: Overall body oval, metallic black in colour; antennae like thread and longer than the head. Anterior feet shorter and the last, most frequently terminated by a compressed tarsus, going in a point. They swim with much swiftness by the assistance of their feet, furnished with long hairs, particularly with last two. All the tarsi have five distinct articulations and of which the anterior two have, in males the first three, articulations very broad and forming altogether a pallet, oval in shape.

Distribution:

Habits and Habitat: Found in ponds and quiet streams.

Good population of this species is found in waters of Khajjiar Lake.

Family: Scarabaeidae

Heavy bodied, oval or elongated, usually convex beetles; tarsi 5 segmented and antennae 8 to 11 segmented and lamellate. The last three of the antennal segments expanded into plate like structure that may be spread apart or united to form a compact terminal club. Front tibiae more or less dilated, with the outer edge toothed.

Genus: *Clinteria* Bwm.

1842. *Clinterla*, Bwm., *Handb. Ent.* iii, p. 299

4. *Clinteria* sp.

Diagnostic Characters: A short and compact insect. Clypeus quadrate and slightly bilobed. Eyes prominent. Outer margins of elytra strongly sinuated and apical angles not acute. Sternum produced between the middle coxae into a short pointed process. Legs not long and three sharp teeth on the front tibia present. Chitinous lobe of mandible long and straight. Maxilla not armed with dense hairs. Last joint of all the palpi large. Males and females slightly different from each other. In males the front tibiae generally a bit more slender and the abdomen longitudinally channelled beneath except in the first group of species.

Distribution: Found in the oriental and Ethiopian Regions.

Genus: *Mimela* Kirby

1844. *Mimela,* Kirby, *Trans. Linn. Soc.* xiv, 1825, p. 101

This genus has various forms generally broadly ovate, sometimes globose. Head generally broad, with short clypeus. Maxilla armed with five very strong parallel teeth in three ranges. Mandible strongly rounded externally and divided at the extremity into two blunt lobes. Elytra have a narrow membranous external margin. Legs short and stout. Front tibia armed with 1 or 2 teeth. The upper tooth present in females, absent in males. The most distinctive feature-the prosternal process, which, behind the front coxae, elevated to their level and angularly produced forward.

5. *Mimela pectoralis*, Blanch 1851

(Plate 9 C)

1851. *Mimela pectoralis,* Blanch. *Cat. Coll. Ent. Mus. Paris.* p. 197

Size: Length: 17 mm, breadth 8 mm.

Diagnostic Characters: Overall colour deep metallic green, with the tibiae and tarsi coppery red. Eyes very large and head coarsely rugose. The clypeus narrow and sub rectangular. The pronotum punctured, with an incomplete smooth median longitudinal line. The lateral margins of prosternum rounded and all the angles not sharp but distinct. The elytra not irregularly but strongly and closely punctured with irregular impressed lines of confluent punctures and a smooth line adjoining the suture. The pygidium scantily punctured. Abdomen smooth in the middle. The prosternal process rounded behind and the mesosternum not produced. The club of the antenna very long.

Distribution: Found in Pakistan, Himachal Pradesh and rest of North India.

Genus: *Onitis* Fabricius

1978. *Onitis* Fabricius, *Ent. Syst. Suppl.,* p. 2

Body oblong; legs not long; front tarsi lacking. Head not broad and the prothorax without process with a pit on each side near the middle. Scutellum minute but visible. Front tibiae armed with four external teeth. Elytra completely cover the abdomen. Front legs elongated. Front legs generally elongated, tibiae slender and strongly curved towards the end. Front tibiae of female always broad with external teeth.

6. *Onitis subopacus* (Arrow, 1931)

(Plate 9 D)

1875. *Onitis philemon* Lansberg, *Ann. Soc. Ent. Belg.* Xviii, p. 133

1931. *Onitis subopacus* Arrow. *Fauna Brit. India, Ceylon & Burma: Col,* iii p. 395

Diagnostic Characters: Overall black with slight metallic lusture; head and prothorax moderately shinning and elytra dull. Clypeus elliptical; pronotum closely punctured with smooth median longitudinal line. Pygidium opaque and scarcely punctured. In males front tibia elongated, slender and strongly curved with a double tooth near the base. The middle tibia slender at the base and strongly dilated. In females front tibia, short and bears four stout external teeth. Middle and hind legs not toothed.

Distribution: In India, found in Kashmir, Assam, Bengal, Bihar, Uttarakhand, Himachal Pradesh and Punjab. Also occurs in Nepal and Burma.

Onitis subopacus can be seen in good numbers in spring and summer seasons in Khajjiar area.

Genus: *Popillia* Serv.

1825. *Popillia,* Serv., *Encycl. Meth.* x, p. 367

7. *Popillia* sp.

Insects of genus, small in size, and short and stout in shape. Head small. Clypeus broadly rounded. Prothorax narrow and dilated at the base. The elytra short exposing the pygidium and part of the propygidium. Legs stout. Front tibia armed with two teeth externally and the tarsi not long. Antennae short and consist of nine joints. The mandibles short and externally rounded. Maxilla armed with five or six very strong sharp teeth. The mentum long and slightly bilobed in front. All the palpi short and stout. Legs of the male thicker than those of the female.

Genus: Melolontha Fabricius

8. Melolontha furcicauda Ancey, 1881

1881. *Melolontha furcicauda* Ancey, *Nafuralisfe* III, p. 412

Size: Length 32.0 mm; breadth 15.0 mm.

Diagnostic Characters: This species has an elongate and convex body, brown-red in colour and covered all over with pale scales. Head closely setose and densely-unevenly punctuate. Clypeus rectangular with strongly reflexed margins and feebly excised in front. Antennae have 10 segments. Club very long in males and 7-segmented. Sides of pronotum finely and densely granulate, closely punctated in middle, with scales minute on sides and large in middle. Scutellum very finely and closely punctate. Abdomen has large sides with white patches of dense fine scales. Pygidium long and projected behind into a bifurcated tail.

Distribution: Found in Himachal Pradesh and Kashmir in India. Also present in China.

Habits and Habitat: Found in association with coniferous forests up to an altitude of about 3000 meters. The adults, on wings during the late half of June and feed. Harmful insects as their grubs damage potato and other vegetable crops. Also feed on roots in seed beds of deodar nurseries and the grubs cut through roots of seedlings and young plants or more often gnaws away the bark all round, thus girdling them. Generally reported from edges of Khajjiar meadow.

Genus: *Anomala*, Samouelle

1819. *Anomola*, Samouelle, *The Entomologist's Useful Companion*, p. 19

Different species of this genus have various forms, some short and globose, others very long and narrow. The surface, smooth and shining, sometimes entirely clothed with hair. The membranous external fringe of the elytra, distinct and considerably developed in some species. The eyes prominent and large. The prothorax transverse slightly lobed. The legs variable in development in different species of the genus. Front tibia armed with one, two, or three teeth. The middle and hind tibiae, generally spinose externally, and each bear two terminal spurs. In the male the apical tooth of the front tibia usually shorter and sharper than in the female, and the inner front claw more or less dilated.

9. *Anomala rufiventris* Redtenbacher, 1842

(Plate 9 E)

1842. *Anomala rufiventris* Redtenbacher, in Hugel's, *Kaschmir, Kafer*, IV, p. 526

Size: Length 16.0-22.0 mm; breadth 10.0-11.0 mm.

Diagnostic Characters: Body elongate-oval in shape; smooth and shining, very dark bronzy-green; sternum and legs greenish-black; ventral abdominal segments mahogany red. Pygidium with few long and erect hairs. Clypeus short and minutely punctated. The forehead with more scattered punctures. Pronotum thinly and very minutely but equally punctuated. Front tibia bidentate, the upper tooth rather slight. The front tarsi and joints claw-joints long. Antennae longer in males than females. Terminal tooth of the front tibia sharp and the inner front claw little dilated and very acute.

Distribution: In India found in Himachal Pradesh, Uttar Pradesh, West Bengal, Manipur, Meghalaya and Sikkim. Also present in Bhutan.

Present in good number in the Khajjiar meadow.

Family: Hydrophilidae

Water scavenger beetles, easily recognized by oval and convex shape, and by the short clubbed antennae and long maxillary palps. Most of the species aquatic and very similar to members of family Dytiscidae. Aquatic species generally black in colour and vary in length from few millimeters to 40 mm. Mesosternum in some species prolonged posteriorly as a sharp spine. Differ from Dytiscids as they rarely hand their head downwards from surface of water.

Genus: *Helochares* Mulsant, 1844

Body shape moderately convex. Pronotum narrowed posteriorly.

10. *Helochares* sp.

Diagnostic Characters: Dorsal surface reddish brown; head darker posteriorly, with well defined paler preocular spots. Labrum reddish yellow with posterior 2/3 darker. Antennae reddish yellow with brownish club. Maxillary palpi reddish yellow. Elytra with incomplete longitudinal rows of dark spots. Dorsal surface shining. Elytra very finely and densely punctured, distance between punctures not larger than diameter of punctures. Head and pronotum with slightly denser punctuation than elytra. Legs reddish yellow with pubescent portion of femora black.

Can be collected from stagnant waters of the Khajjiar lake.

Family: Coccinellidae

Commonly known as ladybird beetles, have brightly coloured small, oval and convex body, distinguished from others in having three tarsal segments. Tarsal claws toothed at the base. Antennae short. Antennae and head often hidden from above. Anterior margin of the pronotum straight. Most species predaceous and very common on vegetation.

Genus: *Coccinella* Linnaeus

General body colour yellowish to reddish brown; head black with two creamish spots in between the eyes. Elytra reddish brown with black spots, ventrally entire black. Pronotum large, punctate, broad at base and slightly narrow anteriorly. Scutellum small, black and triangular.

11. *Coccinella septumpunctata* Linnaeus, 1758

(Plate 9 F)

1758. Coccinella septumpunctata Linnaeus, *Syst. Nat.*, 10th ed., 475

Diagnostic Characters: Body form usually round and oval, yellowish brown in colour. Head convex and deeply sunk in thorax and black in colour. Head covered from above by pronotum; clypeus creamish anteriorly. Eyes large, prominent, well developed and lateral in position towards the base of head. Antennae inserted at the inner front margin of the eyes, black, pubscent and 11 segmented. Scape narrow at the base with broader apex. Three segments form the club and the apical segment globular. Legs short and strong. Elytra convex, hard, punctuate and reddish brown.

Family: Meloidae

Usually narrow and elongated beetles with soft and flexible elytra, and narrower pronotum. Known as bristle beetles because body fluid of the most species contain cantharadin, a substance which cause bristles when applied to skin. Several species of this family important pests.

Genus: *Mylabris*

Colouration of the body variable, usually black with red spots, stripes or barren. Head large, black, broader than pronotum. Antennae 11 segmented, black, short and clubbed; scape swollen with long erect pubescence.

12. *Mylabris pustulata* (Thunberg, 1821)

(Plate 9 G)

Diagnostic Characters: Body thin and elongated; overall colouration black, with red stripes. Head large, black and broader than the pronotum. Clypeus small. Eyes large, bulged and slightly emarginated, placed anteriorly on the lateral sides. Antennae

11 segmented and black in colour. Legs black and long. Tarsi long and five segmented except meta thoracic legs which have four segmented tibiae. Elytra long and less parallel. Each elytron separately rounded at the apex and completely cover the abdomen. Upper surface has three red stripes i.e. antero-lateral, median and posterior, extending from suture to lateral margins. Abdomen large and laterally flattened.

Distribution: In India, found in Tamil Nadu and north India. Elsewhere found in Afghanistan, central Asia, France, Russia, Turkey, Spain and Italy.

Habits and Habitat: Feeds on various types of cultivated and wild vegetation therefore, considered as pest. Generally observed in good numbers in Khajjiar meadow in summer and rainy seasons.

Family: Cerambycidae

Long horned beetles with elongated and cylindrical body, and with long antennae. Antennae usually more than half as long as body of the insect, inserted on frontal prominences, their insertions often surrounded by eyes. First antennal segment 5 times as long as the second. Eyes emarginated, rarely completely divided. The tarsi appear to be 4 segmented with third segment bilobed but actually 5 segmented. Third segment smaller and concealed in the notch of third and difficult to see.

Genus: *Dorysthenes* (*Lophosternus*)

1844. *Lophosternus*, Guer. *Icon. Regm Anim., Ins.* p. 209

Head elongated behind the eyes, short in front. Antennae shorter than the body and 11 jointed. Prothorax transverse and convex, with oblique and denticulate lateral edge in front. Elytra more than twice as long as broad, slightly narrowed behind and rounded at the apex. Legs long and tarsi elongated.

13. *Dorysthenes* (*Lophosternus*) *huegeli* Redtenbacher, 1848

(Plate 9 H)

1848. *Cyrtognathus hugelii*, Redtenbacher. *Hugel's kaschmir*, IV.2. p. 550

1906. *Lophosternus hugelii*, Gahan. *Fauna Brit. India*. Col., 1:11

1979. *Dorysthenes huegeli*, Hayashi. *Ent. Rev. Japan*. 33(1/2): 85

Diagnostic Characters: Overall colour of the body, chestnut-red. Head and prothorax darker than the elytra, the front and hind margins of the pronotum almost black. Eyes large, the upper lobes rather closely approximated to the antennal tubers in front. Antennae little shorter than the body. Pronotum finely and closely punctured in front. Ridges of elytra finely punctured, each with two or three feebly raised obtuse costa.

Distribution: Found in northwest India, Himalayas, Assam, Sikkim, Punjab and Nepal,

Habits and Habitat: A nocturnal insect, found in edges of the forests. This species lives on the edges of the Khajjiar meadow and can be seen especially at night.

A *Calosoma* sp.

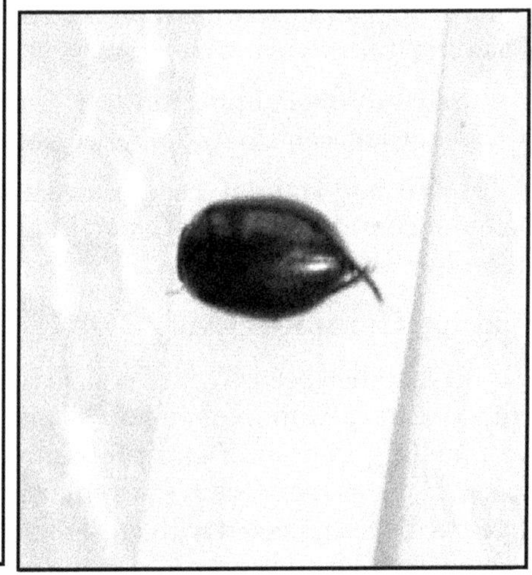

B *Dyticus punctalis*

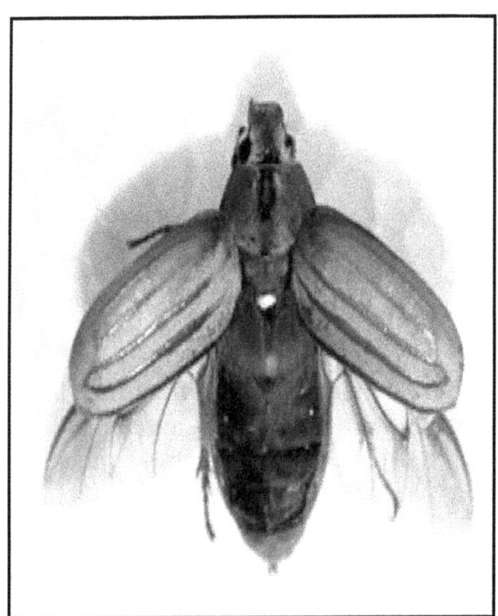

C *Mimela pectoralis*

D *Onitis subopacus*

E *Anomala rufiventris*

F Coccinella septumpunctata

G *Mylabris pustulata*

H *Dorysthenes* (*L.*) *huegeli*

Plate 5.10: Some coleopterans recorded in Khajjiar lake area

Genus: ***Macrotoma*** **Serville**

1832. *Macrotoma,* Serville, *Ann. Soc. Ent. Fr.* i, p. 137

14. *Macrotoma* sp.

Diagnostic Characters: Head elongated behind the eyes; clypeus depressed; eyes not deeply emarginated. Antennae shorter than the body and 11 jointed. Pronotum strongly deflexed at the sides, especially just before the middle. Elytra much more than twice as long as their width across the base and rounded at the apex. Legs rather long and spinose beneath. Tarsi long, with the first joint narrow and much longer than the second.

Distribution: Present in West-African, and Oriental Regions.

Only a few specimens can be seen in Khajjiar pasture during summer months.

Family: Chrysomellidae

Phytophagus, leaf beetles, with variable size, shape and colour of the body, in different species. Antennae generally smaller, clubbed and three segmented. Adult beetles feed on flowers and foliage.

Genus: ***Cleorina*** **Lefev.**

1885. *Cleorina,* Lefev. *Cat. Eumolp.* P. 143

15. *Cleorina* sp.

Diagnostic Characters: Body very round and convex with highly metallic appearance. Head deeply inserted; eyes large and feebly emarginated on the inner orbits. Clypeus not separated from its face. Antennae filiform and the third joint twice as long as second, terminal joints widened. Thorax transverse, convex, and narrow at base and apex, deeply transversely sulcate near the anterior margin. Sides of thorax straight. Elytra wider at the base than the thorax. Legs robust; femora unarmed and claws appendiculate. Prosternum very broad and flat, base truncate. Anterior margin of thoracic episternum convex.

Present in very high numbers in Khajjiar pasture.

F) ORDER: LEPIDOPTERA

Members of this order readily recognised by the scales on wings, which comes off like dust on touching the insect; most of the body and legs are also covered with scales. Mouthparts mostly sucking type. The labrum small and usually in the form of a narrow transverse band across the lower part of the face. The mandibles nearly always lacking. The maxillary palps generally small or lacking but the labial palps always well developed. The compounds relatively large and composed of large number of facets. Order Lepidoptera, divided into two sub-orders, the Rhopalocera and the Heterocera. Butterflies placed in the Rhopalocera while the moths placed in the sub-order Heterocera.

49 species of butterflies harbours Khajjiar belonging to 41 genera and 10 families. Analysis of data reveals that family Nymphalidae and Satyridae (12 species each) dominates the Lepidoptera fauna of Khajjiar area, followed by Pieridae and

Lycaenidae (6 species each), Hesperiidae (4 species), Erycinidae, Papilionidae (3 species) and Danaidae (2 species each), and Acraeidae and Riodinidae (1 species each). Categorization of the species further reveales that of these 49 species, 5 are very common, 32 common, 5 uncommon and 7 are rare. Moreover, 3 species are placed under Wildlife Protection Act (1972). These includea *Lethe scanda* and *Lampides boeticus* placed under scheduled II and *Castalius rosimon* under scheduled IV of the Act.

Family: Papilionidae

Commonly known as Swallowtails, however family includes both tailless and tailed species. Family includes very beautiful, brilliantly coloured species. Predominant colour either black or dark brown in majority of the species. Love visiting flowers or basking in sun shines and fly leisurely over flowers and bushes.

Key to the genera

1. Hind wing with basal cell almost obsolete. Precostal vein short and straight forewing with vein 9 absent..........................*Parnassius*
2. Hind wing with small basal cell; precostal vein directed distal, Green blue discal band on forewing absent..............................*Papilio*

Genus: *Papilio* Linnaeus

1758. Papilio Linnaeus, *Syst. Nat.*, (10th ed.) 1: 458

1. *Papilio protenor* Cramer, 1775 (The Spangle)

(Plate 10 A)

1775. *Papilio protenor*, Cramer *Pap. Exot.*, 1: 77

Wing expanse: 100-130 mm

Diagnostic Characters: Male and female alike, tailless, indigo blur black, duller on forewing than on hind wing in males, paler in females. Hindwings in males with a broad pale yellowish white stripe below upper margin. Underside forewing dull black. Hindwings ground colour as above.

Distribution: Found in Western Himalayas from Kashmir to Kumaon. Elsewhere: Nepal, Bhutan and Bangladesh.

Habits and Habitat: Fond of setting in the company of its own species on damp patches and corners readily to flowers. Prefers wooded hillsides upto 2,590 m, keeps to underground. Flight leisurely, erratic and rapid when alarmed. On the wings from March to September. It is a common species present in Khajjiar area.

2. *Papilio polyctor polyctor* Boisduval, 1836 (The Common peacock)

(Plate 10 B)

1836. *Papilio polyctor polyctor* Boisduval, *Spec. Lep.* 1: 205

Wing expanse: 90-130 mm

Diagnostic Characters: Sexes very nearly alike, mild sight difference between the spring (dry season) and summer (wet season) broods. Male with dark brown scent stripes on forewings. Predominantly black, thickly with golden green scales. Forewings with the band, broader in females and wet season form. Hind wings becoming bluish anteriorly; discal patch blackish blue. Upperside chocolate brown to blackish brown. Hindwings without the characteristic discal patch, a series of claret-ret lunules before the outer margin prominent.

Distribution: Found along the Himalayas from Jammu and Kashmir to Arunachal Pradesh. Also found in Nepal, Bhutan, Myanmar and Pakistan.

Habits and Habitat: Prefers forest Himalayan hills and warm valleys from 600 to 2,100 m. On wings from March to October; comes to flowers. Males fond of settling on wet area. Can be observed in spring, summer and monsoon months in Khajjiar area.

Genus: *Parnassius* Latreille

1804. *Paranassius* Latreille, *Nouv. Dict. Hist. Nat*, 24:185, 199

3. *Parnassius hardkwickei hardwickei* Gray, 1831 (The Common Blue Apollo)

1831. *Parnassius hardkwickei* Gray, *Zool. Misc.*, 1: 32

Wing expanse: 50-65 mm

Diagnostic Characters: Male: tailless, creamy white. Upper side: forewings at base dusted with black thin covering, margin dusky black; area 1 with black edged deep red spot. Hind wing base and the inner margin dusky black; a deep red coloured spot each at base of vein 5 and 7. Underside white and thin. Hind wings spots deep red, ringed black and those in area 2, 5 and 7 centered with white. Female: tailless; dusky black on upper side; underside paler, marking large and prominent; all the red spots centered with white.

Distribution: Found from Jammu and Kashmir to Sikkim. Also occurs in Pakistan and Nepal.

Habits and Habitat: Flies close to the ground in the sunshine and frequently settles on flowers to feed. Seen at heights between 3,000 to 3,350. Not a common species in Khajjiar area. Generally observed in month of May to September.

Family: Pieridae

Commonly known as whites or yellows due to predominant white or yellow colour. Size varies from 25 mm to 100 mm. Some members greatly affected by the season, which create dry season form in winters and wet season form in rainy months. Large numbers of individuals congregate on damp grounds and often accompanied by members of same family or Papilionids. Flight usually slow and close to ground.

Key to the genera

1. Forewing vein R_2 absent..*Delias*
2. Fore wing with vein R_2 arising from the cell...............*Eurema*
3. Forewing vein R_3 emitted from R_4+R_5 very close to apex; stalk long..*Pieris*

4. Hind wing produced a sharp tooth at vein 4. Palpi with short hair..*Gonepteryx*
5. Hindwing with precostal vein short..................*Catopsilia*

Genus: *Delias* Huebner

1819. *Delias* Huebner, *Verz. bekannt. Schmett*, 6: 91

4. *Delias belladonna horsfieldi* (Gray, 1831) (The Hill Jezebel)

1831. *Pieris horsfieldi* Gray, *Zool. Misc.*, 1: 32.

Wing expanse: 70-96 mm.

Diagnostic Characters: The ground colour black; both wings with variable white to cream yellow spots and invariably dusted with dark thin coverings. Upper forewing with elongated areas, spot at the end white; between middle of wing an outer margin and slightly middle of the outer margin series of white spots. Upper hindwing with a large based yellow spot; a tornal yellow patch and whitish inner area. Underside forewings an elongated area stripe wide and whitish.

Distribution: Found from Kullu to Kumaon, also Sikkim. Elsewhere found in Nepal, Bhutan and Myanmar.

Habits and Habitat: Male flies close to the ground. Flight slow but can fly fast when alarmed. Reported from 600 to 3,000 m. Male often seen on flowers. Generally on wings from April to July and September to November; Present in open areas around fields, forest edges and human habitations in Khajjiar area.

Genus: *Pieris* Schrank

1801. *Pieris* Schrank, *Fauna Bioc.*, 2 (1): 152, 161

5. *Pieris canidia indica* Evans, 1926 (The Small Cabbage White)

(Plate 10 C)

1926. *Pieris canidia indica* Evans, *J. Bombay nat. Hist. Soc.*, 31: 312

Wing expanse: 45-55 mm.

Diagnostic Characters: Upperside of forewing at tip and outer margin, usually black up to veins 3 in male and 2 in female, the inner edge of which is dentate; a prominent black spot in area 1b. Hindwing's upper margin with a black spot; terminal black spots at vein endings 3 to 6 in male, but upwards from vein 2 and better developed in female. Underside of Forewing without black wing tip area and marginal band; the black spot in area 3 and 1b large and prominent; upper margin dusted with some black thin covering. Hindwing strongly dusted with black thin covering nearly throughout, heavily at costa, leaving a yellowish area with a large stripe from base through cell and area 4 up to outer margin.

Distribution: In India, found from Jammu & Kashmir to Arunachal Pradesh, hills of northeast, and hills of south India. Elsewhere present in Pakistan, Nepal, Afghanistan, Bhutan, Bangladesh and Myanmar

Habits and Habitat: Mainly a hilly species but migrates to lower elevations and adjoining Himalayas during extreme cold weather. Flight stronger than that of other genus of the family. In Himalayas, seen from 1,500 to 3,600 m. Can be seen to fly low in open places around bushes and shrubs in Khajjiar area. Comes to flowers and damp patches on hot and dry days.

Genus: *Catopsilia* Huebner

1819. *Catopsilia* Huebner *Verz. bekannt. Schmett*, (6): 98

6. *Catopsilia crocale* Cramer, 1775 (The Common Emigrant)

(Plate 10 D)

1775. *Papilio crocale* Cramer, *Pap. Evot.*, 1: 87

Wing expanse: 55-75 mm.

Diagnostic Characters: Male: Upperside, overall colour chalky-white, with either proximal yellow areas or entirely suffused with yellow. Forewing with costa narrowly black upto the base; outer margin with narrow black border, wider at the apex and narrowing posteriorly upto about vein 2 or sometimes to the tornus. Hindwing unmarked, rarely with marginal black vein dots. Underside without markings, yellow or yellowish-white. Forewing often white in the posterior distal area. Female: Upperside creamy white to yellow. Forewing with a black discocellular spot, usually forming a bar to the costa; costal margin black from base to apex; an outer marginal black border, wide at the apex, reaching the tornus, Hindwing with outer marginal blacker border, more or less dentate on its inner edge; a blackish submarginal line, more or less heavily marked, and usually formed of more or less distinct lunules, cutting off submarginal spots of the ground color.

Distribution: Found throughout India. Elsewhere occurs in Sri Lanka, Myanmar, and China.

Habits and Habitat: Recorded up to 2,400 m in the Himalayas. Mainly found during May-October, generally an autumn flying insect; a fast flier, rises high in the air and can cover long distance; flight straight, and powerful.

Can be seen congregating on damp patches on the roadside, forests and visiting some flowers in Khajjiar area. Shows huge emergence during the monsoon season.

Genus: *Gonepteryx* Leach

1815. *Gonepteryx* Leach, *Edinburgh Encycl*; (10th ed): 128

7. *Gonepteryx rhamni nepalensis* Doubleday, 1847 (The Common Brimstone)

(Plate 10 E and F)

1847. *Gonepteryx nepalensis* Doubleday, *Gen. Diur. Lep.* 1: 71

Wing expanse: 60-70 mm

Diagnostic Characters: Head with antennae, palpi and thorax reddish-brown above, thorax with dull yellowish grey hairs; abdomen black above yellow laterally, underside yellowish white. Male, above sulphur yellow. Female, creamy white. Male and female, costa rounded before the apex. Forewing, apex produced and termen below it falcate. Hindwing termen not crinkled. Toothed at vein 3. A dark orange spot at the end of the cell of each wing.

Distribution: Found from Kashmir to Northeast India. Elsewhere: Myanmar, Nepal, Pakistan and Bhutan.

Habits and Habitat: A common species, flies fairly strongly and visits the flowers. On wings from late March to October, between 1,200 and 2,400 m.

Generally present over open patches of grasses in well wooded areas in Khajjiar area. Sometimes, also recorded at damp patches.

Genus: *Eurema* Hubner

1819. *Eurema* Hubner, *Verz. Bekannt. Schmett.*: 96

8. *Eurema hecabe fimbriata* (Wallace, 1867) (The Common Grass Yellow)

(Plate 10 G)

1867. *Terias fimbriata* Wallace, *Trans. Ent. Soc. Lond*, 4 (3): 323

Wing expanse: 30-40 mm

Diagnostic Characters: Head with yellowish antennae, the club black; head, thorax and abdomen yellowish, shaded with fuscous. Underside yellowish white. In dry reason from male's upper side of both wings dark yellow. Forewing with tip broadly black, the inner edge of black border excavated slightly or deeply between veins 4 and 2. Under side usually paler than half of upper side. Forewing with a prominent reddish brown patch; two small spots or specks in basal half on the both sides of median vein, from base to vein 2. Hindwings with a slightly curved sub based series of three spot; discocellular spots or ring irregular curved lines. In dry season form, male and female nearly similar to wet season form in coloration, except that the black border, on both wings, narrower. Underside marking prominent reddish brown.

Distribution: Found in peninsular India from Punjab to Kumaon. Elsewhere: Nepal, Pakistan, Bangladesh and Bhutan.

Habits and Habitat: One of the common among Indian butterflies, visits flowers and damp patches. Known to migrate in large numbers. On wings from April to November. This is a common species, from open areas in Khajjiar madow.

Genus: *Colias* Fabricius

1807. *Colias* Fabricius, *Mag. F. Insektenk (Illiger)* 6: 284

9. *Colias electo fieldi* Menetries, 1855 (The Dark Clouded Yellow)

(Plate 10 H)

1855. *Colias fieldi* Menetries, *Enum. Corp. Anim. Mus. Petr.*, 1: 79

Wing expanse: 42-45 mm.

Diagnostic Characters: Male: Upperside deep cadmium orange-yellow. Forewing with a patch of greenish-black scales at extreme base; a discocellular black pear-shaped spot; a broad outer black border, its inner edge curved slightly and irregularly crenulate, broader at apex and tornus than in middle. Hindwings at the base with a thin covering of long soft hair, beneath which is a dusting of black scales that is continued outwards along the posterior part of the wing. Underside light orange-yellow, the costal margin narrow, outer one fourth part of forewing and whole Hindwing overlaid with pale dull green; costa, outer and inner margins, with cilia of both wings, salmon-pink. Forewing with discocellular spot as on upperside, but with silvery white centre. Female: Upperside with basal black dusting more extensive than in the male, especially on Hindwing. Forewing with submarginal bright yellow spots. Hindwing with discocellular spot, without the central dark rings, but conspicuous in male, Underside as in the male.

Distribution: In India, found in Punjab, northwest Himalayas and Sikkim. Elsewhere occurs in Pakistan and Myanmar.

Habits and Habitat: Widespread and very common throughout Himalayas, seen from February to November, between 900-4,500 m. Flight swift and straight, often settling with wings closed when feeding or resting. A very common species in Khajjiar area.

Family: Danaidae

Generally of moderate to large size; slow and sluggish fliers. Possess an unpleasant smell and acrid juices, which give them protection from their natural enemies. Many of the member species have the habit of congregating on a single plant, sometimes in thousands. Tough butterflies and need prolonged pressure at thorax while killing them. Popular as Tigers and Crows.

Key to the genera

1. Hindwing cross veins m_1-m_2 and m_2-m_3 forming angle .... *Parantica*
2. Hindwing submarginal series of spots evenly aligned and close to marginals .... *Danaus*

Genus: *Danaus* Kluk

1780. *Danaus* Kluk, *Zwierz. Hist. Nat. Poez. Gospod.*, 4: 83

10. *Danaus genutia* (Cramer, 1779) (The Common Tiger)

(Plate 11 A)

1779. *Papilio genutia* Cramer, *Pap. Exot.*, 3: 23

Wing expanse: 70-78 mm.

Diagnostic Characters: Upper side of forewing brownish yellow, with the veins crossing it blackened; apical half black extending along upper margin to base and inner margin to hind angle; a band of white elongated spots before outer margin and at the terminal edge. Hindwing paler, with veins broadly bordered with black; outer margin narrowly black, bearing two more or less complete series of white spots. Underside of forewing with an apical area dusky brown. Hindwing paler; the white spots distinct.

Distribution: Throughout India. Elsewhere: Afghanistan, Myanmar, Japan, Nepal, Pakistan, Sri Lanka, South East China, Thailand and Vietnam.

Habits and Habitat: Flight slowly, close to the ground. Found in highly scruby forests throughout year and abundant in area with heavy rainfall. When occurring in drier region, the tawny part of hindwing tends to become whitish. Found up to 2,400 m in the Himalayas. Emerge in huge numbers in the summer in the Khajjiar area.

Genus: *Parantica* Moore

1880. *Parantica* Moore, *Lep. Ceylon*, 1 (1): 7

11. *Parantica sita sita* (Kollar, 1844) (The Chestnut Tiger)

1844. *Danais sita* Kollar, In Hugel's *Kaschmir und das Reich der Siek,* 4: 424

Wing expanse: 85-105 mm

Diagnostic Characters: Antennae black, head and thorax spotted with white; abdomen brown to bright reddish yellow above, white below. Upper side of forewing black with blush white semitransparent spots and lines. Hindwings chesnut red, with semi-transparent spots and lines; and elongated area completely filled with broad stripe.

Distribution: Common through India and found to about 5000 m in Himalayas. Elsewhere found in Myanmar, Yunnan, Tibet, China and Thailand.

Habits and Habitat: In western Himalayas this species have four broods. Common in status, flight is slow and it visits flowers. Frequently wooded areas. This is a common species of Khajjiar area.

Family: Satyridae

Species of this family, predominantly light or dark brown coloured, so known as Browns. Majority of them weak fliers, with slow, jerky or bouncing movements; fly close to ground littered with dead fallen leaves, over the meadows and upon grassy slopes of the hills. Most of them don't visit flowers and prefer shady areas. Very often observed to settle on roads and slopes.

Key to the genera

1. Forewing with vein 1A+2A swollen at base...........*Mycalesis*
- Forewing with vein 1A+2A not swollen at base...................2
2. Hindwing outer margin not evenly rounded and never symmetrical about a central axis ....*Lethe*
3. Forewing with vein R_2 alwayas arising from vein R_5.......... *Ypthima*
4. Forewing with vein 1A+2A terminating on outer margin in both male and female...*Melanitis*
5. Forewing with vein 5 and 6 not arising near together at origin ....*Lasiommata*

6. Upperside black. Fore wing of male with a band, wings crossed by a discal white and pale yellow band......................*Aulocera*
7. Forewing with 1dc excurved. Vein 10 from the base

....*Callerebia*

Genus: *Mycalesis* Huebner

1818. *Mycalesis* Huebner, *Zutr. Samme Exot. Schmett;* 1:17

12. *Mycalesis perseus blasius* (Fabricius, 1798) (The Common Bushbrown)

1798. Papilio *blasius* Fabricius, *Ent. Syst.* Suppl.:426.

Wing expanse: 38-55 mm.

Diagnostic Characters: Antennae brown above, greyish white beneath, head, thorax and abdomen brown above, pale beneath. Upperside, both wings pale to dark brown; slightly middle of outer margin and marginal lines closer to each other and paler. Forewing with a white centred yellow-ringed black eye in area 2, rarely a small spot in area 5. Hindwings usefully pale to dark brown, without spot. Under side of both wings with the ground colour as on upper side but crossed by a narrow middle white to bluish-white line, with the inner one edged with dark brown, slightly middle of outer margin and marginal lines paler and more distinct than upper side.

Distribution: Found in Himalayas, from Himachal to northeast India. Also found in Myanmar.

Habits and Habitat: Jerky low flight among and undergrowth. Active mainly during morning & evening. Recorded as a very common species in Khajjiar area.

Genus: *Lethe* Huebner

1819. *Lethe* Huebner, *Verz. Bekannt. Schmett;* (4): 56.

13. *Lethe insane insane* (Kollar, 1844) (The Common Forester)

1844. *Satyrus insane* Kollar, In Hugel's *Kaschmir und das Reich der Siek,* 4 (2): 448

Wing expanse: 55-60 mm.

Diagnostic Characters: Antennae, head, thorax and abdomen all brown. Male, upperside, forewings grayish to dark brown, with bands on underside showing through and two reddish yellow minute spots near the wing tip. Hind wings coloured as the forewings; a post middle curved series of four rounded black spots. Females have post middle broad oblique white band reaching hind angle in forewing near wing tip, spots whitish. Underside, whitish band edged on inner side by irregularly shaded and outer side by dark brown triangular shaded areas; outer marginal areas with three eyes as on forewings. Hindwings, as in male.

Distribution: Found in western Himalayas from Chamba of Himachal Pradesh to Kumaon area of Uttakhand.

Habits and Habitat: The species is rare in nature. It flies low, settles often. Can be seen by the road side of forests along mud banks. Flies between 1,500 to 2,700 m from

May to October. Comes to tree sap and overripe fruits. This is a rare species in Khajjiar area and mainly seen during summer and monsoon months.

14. *Lethe scanda* (Moore, 1857) (The Blue Forester)

1857. *Lethe scandal* Moore, *Lepidoptera Indica*, vols.I-VII, Lovell and Reeve London.

Wing expanse: 55-65 mm

Diagnostic Characters: Shows sexual dimorphism. Males with rich deep purplish-blue colour, discal region paler, outer diffused on margins and dark at wing bases. Upper hindwing has indistinct sub marginal black spots. Female dark brown with two narrow, indistinct yellowish subapical marks an upper forewing. A row of four small black subs marginal spots an upper hind wing. Both sexes reddish brown on under side.

Distribution: In India, found along the Himalayas from Jammu and Kashmir to Sikkim and Arunachal Pradesh. Elsewhere found in Bhutan and Myanmar.

Habits and Habitat: Flies in the hills between 900 and 2,700 m in June, August and September. Female is seldom seen. This is a common species in Khajjiar meadow.

15. *Lethe verma verma* (Kollar, 1848) (The Straight-banded Treebrown)

(Plate 11 B)

1848. *Satyrus verma* Kollar, In Hugel's *Kaschmir und das Reich der Siek*, 4 (2): 447

Wing expanse: 55-60 mm

Diagnostic Characters: Upper side of both wings with narrow, pale, slightly middle of outer marginal lines separated by a black line. Forewings brown, oblique middle broad white band form middle of upper margin to above vein 2 in male and below vein 2 in female. Hind wings brown; generally unspotted but the eyes an underside sometimes showings through as white cantered black eyes. Underside of both wings light brown. Forewings with white band as on above; a pale near wing tip area, with two white centred reddish black eyes in area 4 and 5. Hind wings with a regular lila c-grey lines across middle of an elongated area, from upper margin to inner margin.

Distribution: Found in western Himalayas from Kashmir to Kumoan including Nepal, Bhutan and Myanmar.

Habits and Habitat: Flies close to the ground and often settles on road and paths. Comes to damp patches and overripe fruit. This can be seen in grasses in Khajjiar area during.

Genus: *Lasiommata* Westwood

1841. *Lasiommata* Westwood, *British butterflies and their transformation*, 1: 65

16. *Lasiommata schakra schakra* (Kollar, 1848) (The Common Wall)

(Plate 11 C)

1848. *Satyrus schakra* Kollar, In Hugel's *Kaschmir und das Reich der Siek*, 4 (2): 446

Wing expanse: 45-60 mm.

Diagnostic Characters: Male, upperside, both the wings silky, dark smoky-brown. Forewing with a band of four brownish yellow spots in outer margins. Hindwings paler with a post middle row of six white centred orange ringed black eyes, underside pale grayish white. Forewings with two dark lines crossing the eyes, middle area paler than that of upper side edged outwardly by dark lines; an oblique irregular post discocellular orange brown lines, not reaching hind angle. Female resembles the male but with an additional fulvous and bordering eye outwardly and fulvous streak inwardly, reaching up to hind angle in forewings.

Distribution: Found in western Himalayas, in Jammu & Kashmir to Uttarakhand; also found in Baluchistan, Chitral and Kumaon.

Habits and Habitat: Quite variable in ground colour and common in status. Flight lively, close to the ground. Often settles on rocks and on roads. Fond of flowers. Found between 1,830 and 2,700 m. Present mostly in grassy areas around the Khajjiar lake.

Genus: *Aulocera* Butler

1867. *Aulocera* Butler, *Ent.mo mag.*, 4:121

17. *Aulocera swaha swaha* (Kollar, 1844) (The Common Satyr)

1844. *Satyrus swaha* Kollar, In Hugel's *Kaschmir und das Reich der Siek,* 4 (2): 444

Wing expanse: 60-75 mm.

Diagnostic Characters: Male and female morphologically similar; upper side, deep brown with bronzy sheen. Both wings with middle band of whitish larges spots, of these those in area 3 and 4 conical outwardly; sub apical eyes enclosed by a white spot each on upper and inner side, later not reaching to the upper marginal edges. Hindwings much paler; the middle band broader, narrowing posteriorly but not reaching the inner margin; band not excurved outwardly. Underside much paler, with grayish white striae on forewings. Hindwings with basal area coloured with greenish; outer area with a band of greenish white striae; post middle with spreaded black moon like band; marginal black lines dark in both wings.

Distribution: Found in Himalayas, from Kashmir to Sikkim. Also occurs in Chitral in Pakistan.

Habits and Habitat: Sun loving species, weak flier, found between 1,800 and 3,000 m. More associated with forest areas. It is a common species in Khajjiar area.

18. *Aulocera saraswati saraswati* (Kollar, 1844) (The Striated Satyr)

(Plate 11 D)

1844. *Satyrus saraswati* Kollar, In Hugel's *Kaschmir und das Reich der Siek,* 4 (2): 445

Wing expanse: 60-75 mm.

Diagnostic Characters: Upper side deep brown, with copper like brightness. Forewings with the upper margin arched and the outer margin rounded; middle band of white larger spots from inner margins to area A; larger and rounded eye near the wing tip, enclosed by outer, upper and inner white spots, the last larger in female than in male. Hindwing with the band of very broad white spots. Under side much paler, coloured with grayish white to pale reddish yellow; lines dark brown throughout.

Distribution: Found in Himalayas, ranges from Chitral to Sikkim. From Himachal Pradesh it has been recorded from Chamba, Kangra, Kullu and Shimla district. Also found in Nepal and Pakistan.

Habits and Habitat: A strong flier, basks with closed wings, tilting away from sun. Flies in open grassy meadows from 1,200 to 1,800 m. A common species mostly in grassy areas around the Khajjiar lake.

Genus: *Callerebia* Butler

1867. *Callerebia* Butler, *Ann. Mag. Nat. Hist.* (3) 20: 217

19. *Callerebia annade* (Moore, 1857) (The Ringed Argus)

1857. *Callerebia annade*, Moore, *Ann. Mag. Nat. Hist.* p 226.

Wing expanse: 55-70 mm.

Diagnostic Characters: Upperside dark velvety-brown with rounded wings and strongly arched forewing costa. Upper forewing apex with prominent double-pupilled, yellow ringed eyespot. Differs from others of this genus in having hindwing produced to a rounded lobe in tornal area, and a broad discal band on underhind wing extending into tornal area having two small eyespots. On under hindwing discal area has fine white striations, bordered by obscure submarginal and discal bands. Under forewing deep red-brown at apex and ochreous brown at base.

Distribution: In India found from Jammu and Kashmir to Arunachal Pradesh. Also occurs in Nepal, Bhutan, Myanmar and Pakistan.

Habits and Habitat: Hopping flight, but faster than rest of Arguses. Prefers open fairly drier, rocky hillside from 1,500 to 2,400 m. Can be observed easily, mostly from April to October.

Genus: *Ypthima* Huebner

1818. *Ypthima* Huebner, *Zutr. z. Samme. Exot. Schmett.* 1: 17

20. *Ypthima nareda nareda* (Kollar, 1844) (The Large Threering)

(Plate 11 E)

1844. *Satyrus nareda* Kollar, *In* Hugel's *Kaschmir und das Reich der Siek,* 4 (2): 45

Wing expanse: 30-32 mm

Diagnostic Characters: Male and female, similar in appearance; upperside pale-brown; cilia of both wings whitish-brown, with an anticiliary dark line; the broad submarginal band of underside shows through. Forewing with the usual subapical ocellus. Hindwing with a subtornal and sometimes a minute tonal ocellus. Underside pale ochraceous, thickly marked with short dark brown striae, evenly and uniformly spread; ocelli as on upperside, but tonal one bipupilled and always present; hindwing also with a large subapical ocellus. Both wings with sub-marginal, somewhat obscure, dark bands, that on the forewing broadening posteriorly.

Distribution: Found in Himalayas in suitable habitat types from Kashmir to Arunachal Pradesh. Elsewhere present in Nepal, Bhutan, Myanmar and Pakistan,

Habits and Habitat: Comparatively stronger on the wings, flies close to the ground. Usually lurk in undergrowth and bushes and are not often seen abroad; found on grassy river, open and highly wooded forests. Seen on wings from 6,66 to 2,700 m during April to October. A common species, mostly in grassy areas around the Khajjiar lake.

21. *Ypthima ceylonica hubneri* Kirby, 1871 (The Common Fourring)

(Plate 11 F)

1871. *Ypthima hubneri* Kirby, *Syn. Cat. Diurn. Lep.* 95

Wing expanse: 30-40 mm

Diagnostic Characters: Wet-season form, upperside greyish brown. Hindwing usually with two uni-pupilled post-discal ocelli, sometimes with three, rarely all absent. Forewing with subapical ocellus comparatively large, black, bipupilled, with yellow ring. Underside greyish- white, not very densely covered with Short Brown Striae. Hindwing with one apical and typically three posterior post-discal ocelli placed in a curve; traces of discal and submarginal brown bands in most specimens. Forewing with subapical ocellus as on upperside; discal and submarginal obscure dull brown bands; ocellus with a narrow brown ring diffusely produced posteriorly.

Distribution: Found throughout Peninsular India upto Assam. Elsewhere occurs in Myanmar, Malaya and Nepal.

Habits and Habitat: Flutters close to the ground at road sides and grassy land. Reported upto 2,700 m in the Himalayas. Found in wings throughout the year. This is an uncommon species, recorded mainly on the edges of the Khajjiar pasture.

22. *Ypthima sakra nikaea* Moore, 1874 (The Himalayan Fivering)

(Plate 11 G)

1874. *Ypthima nikaea* Moore, *Proc. Zool. Soc. Lond.*, 567

Wing expanse: 45-55 mm

Diagnostic Characters: Upperside, both wings brown, margins darker. Hindwing with bluish-white pupilled ocelli, of which one before apical area; four before outer margin always prominent and distinct than others which may become obscure or completely absent. Forewing with bluish-white, yellow ringed dark brown ocellus, with outer dark brown ring. Underside paler; both wing margins darker; striated with dark brown fine lines. Hindwing with a row of bluish-white yellow ringed eye spots in area before outer margin and two in area before apex, with the upper one smaller and both joined together by a yellow ring in between and both shifted inwards; ocelli in posterior portion, the one in area 2 largest and not in line with those above and below it, the latter bi-pupilled. Forewing with an ocellus before apex larger and distinct than on upperside

Distribution: Found in Bengal, Sikkim and Kumaon. Elsewhere reported from Nepal, Myanmar, Bhutan and Pakistan.

Habits and Habitat: Very common butterfly in western Himalayas. Flies between 900 to 2,700 m, throughout the year. A weak flier, flies close to the ground, seen in light as well as thick forests among cleared undergrowth and at roadsides. Present mostly around grassy areas in Khajjiar pasture.

Genus: *Melanitis* Fabricius

1807. *Melanitis* Fabricius, *Mag. f. Insektenk (Illiger)*, 6: 282

23. *Melanitis leda ismene* (Cramer, 1775) (The Common Evening Brown)

(Plate 11 H)

1775. *Papilio ismene* Cramer, *Pap. Exot.* 1: 41

1994. *Melanitis leda* Varshney, *Oriental Insects* 28: 152

Wing expanse: 60-80 mm

Diagnostic Characters: Wet-season form, upperside brown to dark brown. Forewing with the apex normal and the outer margin angulate or straight; with two black spots before apex, each with a white spot on it; upper margin paler. Hindwing with a prominent white centred fulvous ringed ocellus in hind angle area; some other ocelli and an apical ocellus showing through from below. Underside paler, densely irrovated with dark brown fine lines; yellow ringed with a row of four ocelli on forewing; with a row of six ocelli on hindwing. Dry-season form, upperside rich brown pale or dull. Forewing tip more or less falcate; produced into a tooth at vein 5, the margin straight or sinuous; a large black ocellus before apex, Hindwing brown, toothed at veins 4. Underside greatly variable, with various shades of grey-black, yellow, ochraceous brown and red; ocelli completely obscure on both wings, but when present represented only by white specks or dots. In extreme dry season forms, blotches, bands, lines and striae and an endless variety of patterns are rich, resembling closely to those on dead and dry leaves, where these often settle.

The species shows best example of seasonal changes in the shape of wings, colour variations, particularly on underside, and the markings, especially the ocelli. In dry-season forms, these vary, so much that it is difficult to generalize the description. It will be appropriate to say that no two examples of extreme dry-season match each other in underside wing markings.

Distribution: Occurs throughout India. Elsewhere found in Myanmar, Borneo, Japan, Malaya, Sri Lanka, Sumatra and Taiwan.

Habits and Habitat: One of the commonest of the Indian butterflies. During day time, shelters in undergrowth or among bushes or the roots of trees or even in walkway anywhere that will give it refuge. At dusk it comes out to fly and dances about in the air in a rapid, jerky way. It flies close the ground with its closed wings bent over to one side. This species has been reported upto 2,100 m in the Himalayas. Present mostly around the lake.

Family: Nymphalidae

Brightly and brilliantly coloured butterflies, colour varies from tawny to yellowish and greenish to dark brownish. Usually medium sized to large butterflies. Most species have a reduced pair of forelegs and many hold their colourful wings, flat when resting. Wings show a great variety of pattern formed by bands, spots, patches, stripes, blotches etc. Also called brush-footed butterflies or four-footed butterflies. Underside of the wings often dull and in some species look remarkably like dead leaves, or much paler.

Key to the genera

1. Labial palpi broad, with a dense clothing of scales.........*Athyma*

- Labial palpi narrow, clothing loose scales with numerous fine hairs..3

2. Hindwing of male, vein Sc ends in costa, in female precostal vein with long spur directed distad...................................*Neptis*

3 Forewing pale to dark brown, with the paler half in the tip area and white spots near tip..*Precis*

4. Mid and hind tarsi with patonychia simple..................*Cynthia*

5. Under forewings have sub marginal band composed of black centered white spots..*Parathyma*

6. Three dark vertical lines beyond cell on upper side of both wings, and a row of black spots between the outer two lines

   ....*Pseudergolis*

7. Male and female: fore wings golden brown at base and two-third of inner margin, followed by an oblique red band from upper edge...*Vanessa*

8. Male upperside; forewings deep indigo-blue. Hind wings also deep indigo-blue, with pale blue band in post discocellular area

   ....*Kaniska*

9. Forewings chestnut-red board black bars across middle of an elongated area to near tip area alternating with reddish yellow bars

   ....*Aglais*

10. Forewings reddish yellow, with sinnous black lines, area near wing tip pale reddish yellow; series of blackish spots upto outer margin

    ....*Childrena*

11. Wings pale reddish brown at upper surface, irrorated with darker thin scale at base, upper margin and inner margin; with transverse blackish ....*Issoria*

Genus: *Athyma* Westwood

1850. *Athyma* Westwood, *In* Doubleday's *Gen. Diurn. Lep.*, (2): 272

24. *Athyma opalina* (Kollar, 1844) (The Himalayan Sergeant)

(Plate 12 A)

1844. *Limenitis opalina* Kollar, *In* Hugel's *Kaschmir und das Reich der Siek*, 4 (2): 427

Wing expanse: 55-70 mm.

Diagnostic Characters: Body black above, with an iridescent bluish-white band at the base of abdomen, whitish beneath. Male, upper side, forewings brownish black, with creamy white markings. Hindwings brownish black; middle white band continued from that on forewings, crossed by black veins from upper to inner margin, slightly middle of outer margin series of white moon like from top to inner margin near hind angle. Underside forewing ferruginous, markings as on upper side; hindwings suffused with purplish, a curved line from base to upper margin. Female as the male, but the ground colour on both sides paler and the markings considerably larger.

Distribution: The species occurs from Kashmir to Arunachal Pradesh. Also present in Nepal, Bhutan, Myanmar and Pakistan.

Habits and Habitat: A common species. Not a strong flier compared to outer sergeants. Typical sailing flighty close to the ground. Commonest in sunny nullahas where it beat up and down, often in company of swarms of common and Himalayan Sailors, and settles on damp patches, rocks or on bushes. Seen around forest clearing and paths, between 1,800 and 2,800 m. This is a common species in Khajjiar areas.

25. Athyma *asura asura* (Linnaeus, 1758) (The Studded Sergeant)

1758. *Papilio cardui* Linnaeus, *Syst. Nat.*, 10th ed., 475

Wing expanse: 65-75 mm.

Diagnostic Characters: Male and female same, upper forewings, cell streak entire and narrow with rounded detached spot beyond. Under forewings have sub-marginal band composed of black centered white spots. Unlike other sergeants, it has black centered white spots forming the post discal band on both upper hindwings and under hindwing.

Distribution: Present in India from Himachal Pradesh to Arunachal Pradesh. Also present in Nepal, Bhutan, Bangladesh and south Myanmar.

Habits and Habitat: A robust butterfly, strong on wings, flies along forest streams in sheltered valleys. Fond of visiting damp patches. Seen on wing in August and recorded up to 2,600 m. This species is not so commonly recorded in Khajjiar areas.

Genus: *Neptis* Fabricius

1807. *Neptis* Fabricius, *Mag. f. Insektenk (Illiger)*, 6: 282

26. *Neptis mahendra* Moore, 1872 (The Himalayan Sailer)

1872. *Neptis mahendra* Moore, *Proc. Zool. Soc. Lond.*, 560

Wing expanse: 55-60 mm

Diagnostic Characters: Upperside dark brown to black, markings whitish. Forewing with a single narrow streak, having large conical spot beyond it; an anterior series of white upper marginal streak and large spots; a posterior series of oblique white spots beyond the middle to inner area; a series of smaller white spots before outer margin. Hindwing with a medial band of broad white spots, widening at the upper margin; a series of spots before outer margin from below apex to near hind angle. Underside deep brownish ferruginous; a series of white spots and markings as on upperside but larger and prominent.

Distribution: Reported from Kumaon, Himachal Pradesh and Kashmir in India. Elsewhere found in Pakistan.

Habits and Habitat: One of the commonest butterflies of Himalaya, found from 1,200 to 3,000 m, from April to October. Found commonly in woods, gardens, and does not visit damp places. Sometime seen on the flowers of *Lantana,* for the food. Mostly present on the edges of Khajjiar meadow.

27. *Neptis hylas astola* (Linnaeus, 1758) (The Common Sailer)

(Plate 12 B)

1758. *Papilio hylas* Linnaeus, *Syst. Nat.*, 10th ed. 486

Wing expanse: 50-60 mm

Diagnostic Characters: Black butterfly with white markings; bands and spots broader in dry-season form, and narrow bands and smaller spots in wet-season form. Underside ochraceous in dry-season form and chocolate in wet-season form; dark edges to white bands and spots more prominent. Forewing with a series of white line near wing tip and large spots in upper margin, middle series of white spots and slightly middle of outer margin, a series of small white spots. Hindwings with a middle band of white spots, not widen to upper margin.

Distribution: Found from Kashmir to Arunachal Pradesh. Elsewhere found in Myanmar and Sri Lanka.

Habits and Habitat: A very common species, found everywhere except in the arid regions. Occurs in woods, gardens, damp and forested areas of the hills. Settles on damp ground. Found upto 2,700 m in the Himalayas. It has a beautiful flight, floating in and out of sun light under the shade of trees. Habits and flights typical, after a few flicks of the wings the insect sails gracefully along with wings held horizontal. Observed commonly around the Khajjiar meadow.

28. *Pseudergolis wedah* Kollar, 1844 (The Tabby)

(Plate 12 C)

1844. *Limenitis opalina* Kollar, *In* Hugel's *Kaschmir und das Reich der Siek,* 4 (2): 427

Wing expanse: 55-65 mm

Diagnostic Characters: Both sexes similar. Upper side, golden brown with four dark bars in cells of both wings. Three dark vertical lines beyond cell on upper side of both wings, and a row of black spots between the outer two lines. Underside light chocolate brown with violet gloss. Fore wings apex square cut.

Distribution: Found in Himalayas from Himachal Pradesh to Uttarakhand and from Sikkim to Arunachal Pradesh. Also found in Nepal, Bhutan, Bangladesh and Myanmar.

Habits and Habitat: Flies along streams in hilly areas. Often settles on stones and leaves with wings spread-out flat. Seen on carnivore droppings and damp patches. On wings between 400 to 2,000 m from April to December. Normally seen along the rivulets in Khajjiar meadow.

Genus: *Precis* Huebner

1819. *Precis* Huebner, *Verz .bekannt. Schmett.*, (3): 33

29. *Precis iphita* (Cramer, 1779) (The Chocolate Pansy)

1779. *Papilio iphita* Cramer, *Pap. Exot.*, 3: 209

Wind expanse: 55-65 mm

Diagnostic Characters: Upperside of forewing pale to dark brown, with the apical half paler, white spots before apex; dark-brown obscure markings in the middle and at end of the cell. Hindwing pale to dark brown with the basal half of inner margin whitish; apical half paler, with the markings beyond middle continuing from base on forewing; four to five ocelli present, line before outer margin as on forewing. Underside pale; both wings crossed by markings as on upperside but the one from upper margin near apex to inner angle of hindwing more prominent; ocelli abscure on both wings. Considerable seasonal variations in the species. In dry-season form, the apex of forewing angulated and the hindwing lobed at inner angle, with the underside paler and leaf like.

Distribution: Found from Kashmir to Kumaon. Elsewhere, found in Myanmar, Sri Lanka, Pakistan, Nepal, Bhutan and Bangladesh.

Habits and Habitat: Mostly flies in the hills, upto 2,700 m in the Himalayas, common in wet, well wooded regions and shaded places. On wings from January to December prefers shady places. A very commonly encountered species in Khajjiar area from early spring to late autumn.

Genus: *Cynthia* Fabricius

1807. *Cynthia* Fabricius, *Mag. f. Insektenk (Illiger)*, 6: 281

30. *Cynthia cardui* (Linnaeus, 1758) (The Painted Lady)

(Plate 12 D)

1758. *Papilio cardui* Linnaeus, *Syst. Nat.*, 10th ed. 475

Wind expanse: 55-70 mm

Diagnostic Characters: Upperside of both wings purplish red at base. Forewing dark brown, with pale-ochraeous spot at sub-basal area. Hindwing dark brown, with the area from beyond the cell to outer margin ochraeous brown; a row of larger five dark brown to black spots with slightly paler center, a row of lunular spots before outer margin and a marginal row of larger spots. Underside markings as above but with some additional spots on forewing cell. Hindwing with deeper shade of ochraeous and brown; an oval spot across middle of cell and row of medial spots, as on upperside, but better developed into prominent ocelli, two of the spots being larger with blue centers and black outer ring.

Distribution: Found throughout India. Elsewhere, Myanmar and Sri Lanka.

Habits and Habitat: A common species, frequently settles, most oftenly on the ground or sometimes on a leaf, reported upto 4,500 m in the Himalayas. It delights in fairly open places, such as fields, waste lands, gardens and roads, where it flies strongly and swiftly, in a dashing and discontinuous manner. Widely distributed ranging from higher hills to sea level, probably due to a wide variety of host plants and its migratory habit. Present in open areas of the pasture.

Genus: *Vanessa* Fabricius

1807. *Vanessa* Fabricius, *Ill. Mag*. 6: 281

31. *Vanessa indica* (Herbst, 1794) (The Indian Red Admiral)

1794. *Papilio atlana indica* Herbst, *Nat. Schmett*., **7**: 171

Wing expanse: 55-65 mm.

Diagnostic Characters: Forewings golden-brown at base and two-third of inner margin, followed by an oblique red band from upper edge, through an elongated area and terminating at inner margin before hind angle; hind wings golden brown nearly throughout up to red marginal area. The later with a row of small black spots. Underside paler; forewing tip pale-reddish yellow. Hindwings brown, irrorated with white, grey dark brown to black; marginal area pale reddish yellow.

Distribution: Found in Himalayas and northeast India from Kashmir to Assam. Also present in Pakistan, Afghanistan, Nepal, Bhutan, Myanmar and Sri Lanka.

Habits and Habitat: The species is nearly common throughout India and is widely distributed; can be seen in and around forests and garden too. Frequently settles on ground with wings half open. Can be observed on wings from March to December.

32. *Kaniska canace* (Linnaeus, 1767) (The Blue Admiral)

1767. *Papilio canace* Linnaeus, *Syst. Nat.* 12th ed., 1 (2): 779

Wing expanse: 60-75 mm

Diagnostic Characters: Male upperside, forewings deep indigo-blue; middle band pale blue from area 4 or above, with a few dots at anterior end and widening posteriorly at inner margin. Hind wings also deep indigo-blue, with pale blue band in post discocellular area with a series of black small spots on it; inner margins lighter, outer margins lobed at veins 4. Under side dark and striated, with various shades of grey, black, green, reddish, yellow, or pale violet bands border and prominent, with the edges irregular, spots on bands reddish yellow on forewings and black and hind wings. Female resembles males but larger, with the markings prominent and marginal likes usually obsolete.

Distribution: Found in India from Jammu and Kashmir to Arunachal Pradesh, and hills of northeast and south India. Outside India, present in Sri Lanka, Bhutan, Nepal, Myanmar and Pakistan.

Habits and Habitat: Prefers forested hilly regions. Flies in the vicinity of water, frequently settles on roads and damp patches. Attracted to overripe fruits and tree sap. Seen on the wings from March to November in areas ranging between 1,000 to 3,000 m. Good population of the species is seen in Khajjiar area in summer months.

33. *Aglais cashmirensis* (Kollar, 1848) (The Indian Toroiseshell)

1848. *Vanessa cashmirensis* Kollar, *In* Hugel's *Kaschmir und das Reich der Siek*, 4 (2): 442

Wing expanse: 55-65 mm.

Diagnostic Characters: Upper side, forewings chestnut-red, with broad black bars across middle of an elongated area to near tip area, alternating with reddish yellow bars; a large black patch in inner area followed by a reddish yellow patch and a black spot paler at vein 2. Hind wings with a broad chestnut-red middle band; bases blackish; inner margin paler in basal half; outer margin black followed by two paler lines; margin toothed at vein 4. Under side both wings dusky brown at bases, paler beyond, and lined with black. Shows seasonal variation, wet season form being brighter than those of dry season darker broods.

Distribution: Found in India from Chamba to Sikkim. Also present in Nepal, Bhutan, Pakistan and Afghanistan.

Habits and Habitat: Commonest of the Himalayan butterflies. Seen throughout the year in various habitats except in deep forest. A very common species in Khajjiar area.

34. *Childrena childreni* (Gray, 1831) (The Large Silverstripe)

(Plate 12 E)

1831. *Argynnis childreni* Gray, *Zool. Misc.*, 1: 33

Wing expanse: 75-100 mm.

Diagnostic Characters: Male, upper side, forewings reddish yellow, with sinnous black lines, area near wing tip pale reddish yellow; series of blackish spots upto outer margin. Hind wings as the forewings, inner margins colored with blue; with a line at end of an elongated area; marginal series of black spots, margin dentate. Under side forewings paler; area near tip pale reddish yellow and the tip pale greenish. Hind wings rich reddish yellow-greenish, crossed by numerous silvery lines and bands, edged with black on one or both sides. Female much darker than male.

Distribution: In India found from Jammu and Kashmir to Arunachal Pradesh, and also hills of northeast. Also found in north Myanmar, Pakistan, Nepal and Bhutan.

Habits and Habitat: Common in status, fast flier, often seen along roads, valleys and in open country, settling on damp ground and flowers. Found between 1,200 to 3,000 m in the months of May to November. A commonly seen species in Khajjiar area.

35. *Issoria lathonia* (Linnaeus, 1761) (The Queen of Spain Fritillary)

(Plate 12 F)

1761. *Papilio lathonia* Linnaeus, *Fauna Suec.*: 282

Wing expanse: 55-60 mm.

Diagnostic Characters: Both wings pale reddish brown on upper surface, irrorated with darker thin scales at the base, upper margin and inner margin; with transverse blackish spots followed by middle series of round spots as on forewings. Several salivary patches and spots on under surface of the hind wings.

Distribution: In India found in suitable habitat types from Jammu and Kashmir to Arunachal Pradesh. Also occurs in Nepal, Bhutan north Myanmar, Pakistan and Afghanistan.

Habits and Habitat: A common and also one of the prettiest Indian fritillaries. Fast flier typical butterfly of high attitudes. Found between 1,200 to 5,000 m during the months of February to October. Present in good numbers in Khajjiar area especially in late spring and summer months.

Family: Acracidae

A small family, represented by 2 members only in India. Members small to medium sized with narrow wings and long-slender abdomen. Predominantly, yellow or tawny in colour. Forewings long and hind wings rounded. Flight slow, close to the ground and often settles on flowers, exposed places on leaves, or twigs.

Genus: *Acraea* Fabricius

1807. *Acraea* Fabricius, *Mag. f. Insektenk (Illiger)*, 6: 284

36. *Acraea issoria anomala* Kollar (The Yellow Coster)

1848. *Acraea anomala* Kollar, *In* Hugel's *Kaschmir und das Reich der Siek*, 4 (2): 425

Wing expanse: 45-65 mm.

Diagnostic Characters: Antennae, head, thorax and abdomen black; thorax with a little reddish yellow pubescence anteriorly; underside black but marked with pale reddish yellow spots. Overall appearance fulvous to reddish yellow. Forewing with the veins along upper margin, broadly black and along outer margin narrowly black from tip and traversed by an antemaginal series of smaller spots of ground colour. Hindwings with the ground colouration as of forewings; outer areas with marginal wavy black line, slightly middle of outer margin line wavy. Underside, forewings, yellow, becoming paler towards tip. Hindwings pale and dull but the veins darker; slightly middle of outer margin line lunulate, edged with black on inner as well as outside; marginal line pale.

Distribution: Occurs from Himachal Pradesh to Arunachal Pradesh and northeast India. Elsewhere found in Nepal, Bhutan and Myanmar.

Habits and Habitat: Shows considerable variation especially in ground colour. Flight Slow and fluttering, often settling on leaves and flowers. On the wings from April to September, between 700 and 2,400 metres. Present mostly on edges of the Khajjiar meadow.

Family: Erycinidae

These are popularly known as Beaks, Punches and Judies. Members have predominant colour brown, with bands or spots on wings. Fond of shaded wooded areas, hilly forests in the vicinity of streams.

Genus: *Libythea* Fabricius

1807. *Libythea* Fabricius, *Ill, Mag.*, 6: 284

37. *Libythea myrrha* Godart, 1819 (The Club Beak)

(Plate 12 G)

1819. *Libythea myrrha* Godart. *Enc. Meth.*, 9: 171

Wind expanse: 45-55 mm

Diagnostic Characters: Body stout, thickly covered with soft, dark-cloud wooly hair. Head small, strongly tufted in front. Antennae straight; abdomen short and slender. Male, upperside of both wings deep vinous-brown, almost black, with tawny markings; forewing with a streak commencing narrowly at the base of the wing, occupying the lower half of the discoidal cell and upper half of the sub-median inter-space. Hindwing with a broad discal band and gradually narrowing towards the apex of the wing. Underside much paler; forewing with the discal streak wider, occupying almost the entire basal area of cell, the apex irrorated with purplish. Hindwing without any tawny discal band, irrorated throughout purplish, which assume the form of a more or less distinct band across the disc, and another occupy the space from the middle of the costa to the middle of cell. Female pale throughout, with large tawny markings.

Distribution: Found in Himalayas from Kullu to Assam. Elsewhere found in Myanmar, Java and Borneo.

Habits and Habitat: Often found settling on some dead stick with folded wings. Strong flier, often skipping and sailing. Found between 900 to 3,000 m, from March to October. Recorded mainly in summer months in Khajjiar area.

38. *Libythea lepita* (Moore, 1857) (The Common Beak)

1857. *Libythea lepita* Moore, *Cat. Lep. Mus. E.I.C.*, 1: 240

Wing expanse: 45-50 mm

Diagnostic Characters: Resembles *Libythea myrrha,* however differs by the narrow line in the forewing, separated or connected to a large spot beyond an elongated area by a short neck; the spots near forewing tip area also separate. Hindwing with the middle band short and narrow, and not reaching upper as well as posterior margin.

Distribution: The species ranges from Kashmir to Assam and southern India, besides, found in Sri Lanka, Pakistan, Bhutan, Myanmar and Nepal.

Habits and Habitat: Flight rapidly darting, skipping and sailing. Often settles on damp patches and visit flowers. In Himalayas, flies between 900 to 3,000 m. On wings in Khajjiar area from March to September.

Family: Riodinidae

Known as Metalmarks due to their small metallic-looking spots on their wings. Males have reduced forelegs while the females have full-sized, fully functional forelegs. Hindwings have a unique venation, costa thickened out to the humeral angle and the humeral vein short. Most species perch on the undersides of leaves with the wings held open and completely flat.

Genus: *Dodona* Hewitson

1861. *Dodona* Hewitson, *The Exot. Butt.* 2: 91

39. *Dodona durga* (Kollar, 1848) (The Common Punch)

1848. *Melitea durga* Kollar, *In* Hugel's *Kaschmir und das Reich der Siek*, 4 (2): 441

Wing expanse: 30-40 mm

Diagnostic Characters: Male, upperside of both wings fuscous with numerous bars and spots. Forewing with yellowish short streaks across the middle, at the end and beyond the cell, the latter being the largest yellowish spots present in between veins, below and beyond the cell and before outer margin; marginal row of small five linear spots from below the apex to the inner angle. Hindwing with marks at the end of cell and with ochraeous lines; an irregular medial series of spots and a series of linear spots before outer margin and at the margin which coalesce at anal lobe. Underside of both wings dark ochraeous. Forewing with the base and inner margin fuscous; base of cell enclosing a small black spot; other markings as above. Hindwing with more or less ochraeous markings. Female larger, wings broader with convex outer margin of forewing more convex; tip less protruded, and the markings same as in male. Both male and female tailless.

Distribution: Found from Kashmir to Kumaon. Elsewhere found in Pakistan and Sri Lanka.

Habits and Habitat: Found between 750 to 2,550 m in the Himalayas. Flies in a jerky way and found on the roadside shrubs, settles on the flowers of Asteracea family. Not very common in status and the males fond of visiting flowers and damp patches. Commonly observed from March till October in Khajjiar area, mostly along the edges of the meadow.

Family: Lycaenidae

Mostly blue shades so known as Blues; though suffused, striped, spotted, banded or mottled with a variety of patterns and colours, but devoid of any marks on upper side. Tails at the hind wings, vary from thread like, long or fluffy or even often lobed. When at rest, tails rubbed together with oscillating movements, which resemble the head of an insect or other butterfly with antennae.

Key to the genera

1. In forewing vein R_1 ending well before end cell...........*Castalius*
- In forewing vein R_1 ending opposite end of cell...........*Tarucus*

- In forewing veins R_1 and Sc free; hindwing tailed......*Syntarucus*
- Hindwing tailess...2

2. In forewing vein Cu_{1b} arising opposite to the origin of vein R_1 .... *Chilades*

- In forewing vein Cu_{1b} arising opposite to the origin of vein R_1, near base; hindwing tailed ....*Euchrysops*

3. Underside of hindwing with jeweled spots along margin or at tornus...*Freyeria*

- Underside of hindwing without jeweled spots along margin or at tornus...4

4. Two black tornal spots on upper hind wings..............*Lampides*

- Large well separated black rounded tornal spots outlined with brilliant silvery thin...................................*Pseudozizeeria*

Genus: *Preudozizeeria* Beuret

1955. *Preudozizeeria* Beuret, *Mitt. Ent. Ges. Basel.* 5: 125

40. *Pseudozizeeria maha* (Kollar, 1848) (The Pale Grass Blue)

(Plate 12 H)

1848. *Lycaena maha* Kollar, *In Hugel's Kashmir und das Resch der Siek.*, 4 (2): 422

1994. *Pseudozizeeria maha* Varshney, *Oriental Insects*, 31: 96

Wing expanse: 20-30 mm

Diagnostic Characters: Antennae branded with white, thorax bluish above, abdomen brownish. Underside whitish, covered with white, wooly delicate hairs. Male, upperside of both wings pale greyish-blue; borders narrow and fuscous. Spots occurring underside encircled white. Forewing with a black spot in cell as well as a bar at end of cell; post-discal, sub-marginal and marginal series of black spots, the former not reaching the anterior margin and the spots becoming larger towards posterior margin. Hindwing with three black spots across the wings in upper, median and inner area, a large spot in upper area; large well separated black rounded tornal spots outlined with brilliant silvery thin covering. Female dark brown to fuscous above, with blue scaling. Underside markings more distinct.

Distribution: Reported from West Bengal, Assam and northern India Elsewhere found in Myanmar, Nepal, Pakistan, Bangladesh and Bhutan.

Habits and Habitat: Flight weak, close to ground. Found in Himalayas, upto 3,750 m, on wings from January to November. Prefers open grassy areas, more abundant in hills. Common in grassy areas of the Khajjiar pasture in spring and summer seasons.

Genus: *Lampides* Hubner

1816. *Lampides* Hubner, *Verz. bek. Scumett.*, p. 70

41. *Lampides boeticus* (Linnaeus, 1767) (The Peablue)

1767. *Polyommatus boeticus*, Linnaeus *Syst. Nat. ed.* xii, i, p. 789

Wing expanse: 24-36 mm.

Diagnostic Characters: Underside very pale brown with darker brown bands. No spots on under forewings. Distinct lighter band within outer margin on under hindwings. Two orange ringed black tornal spots, often with metallic silver crown on underside wings. Upper side of male dark violet blue. Two black tornal spots on upper hind wings. Female brown on upper side, overlaid with pale salivary blue scales at wings bases.

Distribution: Found in India, Pakistan, Afghanistan, Nepal, Bhutan, Bangladesh, Sri Lanka and Myanmar.

Habits and Habitat: Common in status, a strong flier, flies close to the ground. Common in agricultural and cultivated areas and in the hills, upto 3,040 m. Visits flowers and damp patches. Known to migrate. Found mainly in grassy areas of the Khajjiar meadow.

Genus: *Lycaena* Fabricius

1807. *Lycaena* Fabricius, *Illig. Mag.* vi, p. 286

42. *Lycaena pavana* (Kollar, 1828) (The White-Bordered Copper)

1828. *Nacaduba pavana* Horsfield (*Lycaena*), *Cat. Lep. Mus. E. I. C.*, p. 77

Wing expanse: 37-40 mm

Diagnostic Characters: Under forewings dull orange. A white band on under hind wings between discal row of spots and sub-marginal double row of spots the distinguishing characteristic. Male reddish-copper on upper forewings with brown border, and two black spots at mid-and-end cell; an irregular discal band of spots. Upper hindwings mostly brown with tinges of copper in cell and basal area, the purplish sheen along margin. Female has entire upper hindwings brown and on upper forewings only discal area and cell orange copper. Similar irregular discal band of spots on upper forewings.

Distribution: Present in India from Jammu and Kashmir to Uttaranchal. Also present in Pakistan and Nepal.

Habits and Habitat: Can be seen among open grassy patches and low flowering plants. Found mainly between 1,700 and 2,800 m from March to October. Generally seen in Khajjiar pasture in summers.

Genus: *Heliophorus* Geyer

1832. *Heliophorus* Geyer, *In* Huebner's *Zutr z. Samml. Exot., Schmett*, 4: 40

43. *Heliophorus sena* Kollar, 1844 (The Sorrel Sapphire)

1844. *Polymmatus sena* Kollar, *Hugel's Kaschmir*, 4: 415

Wing expanse: 28-33 mm.

Diagnostic Characters: Male, upperside of both wings shining violet. Forewing with a well-defined, rather broad outer black margin. Hindwing with the costa and outer margin rather broadly black, the latter bearing a prominent series of orange lunules enclosing rounded black spots. Forewing with a submarginal orange lunulated band from the first median nervule decreasing to the apex. Hindwing with the outer margin similarly marked to that of the forewing but all the markings broader; tail black tipped with white; cilia throughout alternately black and white; body fuscous above, white beneath. Female, upperside of both wings dull fuscous, with no trace of the shining violet coloration present in the male; the orange lunulated submarginal fascia on the hindwing as in the male, but more or less continued on to the forewing towards the anal angle, on the hindwing with a fine blue line following its inner margin, sometimes present in the male. Underside of both wings as in the male.

Distribution: Found throughout India. Elsewhere found in Nepal, Tibet, Pakistan and Afghanistan.

Habits and Habitat: Very common in status and on wings from March to October. Found between 600 to 2,400 m in the Himalayas. Flight not fast, remains around small shrubs for flowers. Flies around low growing vegetation by the road sides, in meadows and waste grounds. Encountered very commonly in Khajjiar area especially during spring and summer seasons.

Genus: *Castalius* Huebner

1819. *Castalius* Huebner, *Verz. Bekannt. Schmett*, 5: 70

44. *Castalius rosimon* (Fabricius, 1819) (The Common Pierrot)

1819. *Castalius rosimon* Fabricius, *Syst. Ent.* 523

Wing expanse: 25-27 mm.

Diagnostic Characters: Male, upper side of both wings white, with a grayish blue base; anterior margin blackish-brown of both wings, a border of same colour passing along the posterior margins, and bearing a regular series of while rings formed by crescents applied to each other. A very irregular series of squares or ablong maculae passes along the inner edge of the marginal border and several elongated maculae. Underside of both wings white. Female larger than male. On the upper side of both wings, the posterior portion broader and black colour more intense. The underside of the both wings resembles that of male.

Distribution: Found throughout India except extreme northwest. Outside India found in Pakistan, Nepal, Bhutan, Bangladesh, Myanmar and Sri Lanka.

Habits and Habitat: Common in status, flight flattering, close to the ground, fond of sunshine, mud puddles, flowers, dead insects and bird droppings. Found in open country as well as in forested region up to 2,500 meters in Himalayas on the wings from January to November. More common during the rains. Present in all habitats of Khajjiar area.

Genus: *Ropala* Moore

1881. *Ropala* Moore, *The lepidoptera of Ceylon*, 1 (3): 105

45. *Ropala manea schistecea* Moore, 1879 (The state Flash)

1879. *Deudorix schistecea* Moore, *Proc. Zool. Soc. London*; 140

Wing expanse: 30-33 mm.

Diagnostic Characters: Male, upper side, both wings dark slaty blue, shot with brilliant blue in lower middle area of hind wings. Hind wings with the inner margin pale fuscous; posterior lobe black, bearing inwardly a patch of dull reddish yellow thin covering, with a patch long white hair above the lobe. Female upper side shining with purplish brown. Seasonal variation in coloration seen.

Distribution: Throughout Indian expect arid zone. Elsewhere found in Sri Lanka, Pakistan, Bangladesh and Myanmar.

Habits and Habitat: Common in status, visit damp patches, flight rapid, seen on both plains and hills up to 1,980 meters. Seen throughout the year but most active in monsoon and post monsoon period. Often seen resting on the under sides of the leaves of trees or bushes. Very fond of flowers of *Lantana*, does not bask in the sun. This species is abundant in all habitats of Khajjiar area.

Family: Hesperiidae

Large butterflies having hairy body with a large head, at least as wide or wider than the thorax; fully developed and functioning forelegs in both sexes and pointed wings. Curved or hooked antennae tips. Known as Skippers, having dark brown to black colouration spotted or banded with white or yellow markings. Males of many species possess a patch of scent scales on the forewing, called a stigma, useful in attracting females. Males may also have a folded portion of the forewing on the leading edge, called a costal fold, which encloses scent scales.

Genus: *Coladenia* Moore

1881. *Coladenia* Moore, *Lep. Ceylon*, 1: 180

46. *Caladenia dan* (Fabricius, 1787) (The Fulvous Pied Flat)

1787. *Hesperia dan* Fabricius, *Mant. Ins.*, 2: 88

1949. *Caladenia dan* Evans, *A catalogue of Hesperiidae from Europe, Asia and Australia in Brit. Mus. Nat. Hist*, 112

Wing expanse: 35-45 mm

Diagnostic Characters: Upper side dark golden brown. Forewing with large spots in cell and below it in area 2; spots semi-transparent white in female, yellow in male, those in area above cell, in area 3 and 1b, small; smaller post-discal spots in 6 to 8. Hindwing with the spots dark and diffused. Underside of forewing with prominent and well defined spots; hindwing markings as on upperside.

Distribution: Found from Kullu to Assam. Elsewhere reported from Myanmar.
Habits and Habitat: Found upto 2,300 m in the Himalayas. Common throughout damper hilly, jungle regions, coming out more often into open areas to bask in sunshine from May to October. This species is generally recorded in forest edges in Khajjiar meadow.

47. *Sarangesa dasahara* (Moore, 1865) (The Common Small Flat)

1865. *Nisoniades dasahara* Moore, *Proc. Zool. Soc. Lond.* 787

1949. *Sarangesa dasahara* Evans, *A catalogue of Hesperiidae from Europe, Asia and Australia in Brit. Mus. Nat. Hist.* 120

Wing expanse: 28 mm

Diagnostic Characters: Upperside dark brown to blackish. Spots on the upperside of forewing minute, sometimes obsolete, in area 2, cell 6 to 8 and in subcostal area, when present, minute. Hindwing spots diffused. Underside greyish brown, with large diffused spots.

Distribution: Found throughout India. Elsewhere: Myanmar and Indo China.

Habits and Habitat: Found upto 2,300 m in the hills. Flies down over grasses and among bushes both in jungle, open places and in hilly regions. Visits flowers of low growing plants, rests on the underside of leaves and bask on their upper surfaces. Present in grassy areas around the Khajjiar lake.

48. *Polytremis eltola* (Hewitson, 1869) (The Yellow-Spot Swift)

1869. *Hesperia eltola* Hewitson, *Exot. Bult.*, 4: pl. 4, fig. 40

1944. *Polybemis eltola* Evans, *A Catalogue of Hesperiidae from Europe, Australia and Asia in Brit. Mus. Nat. Hist.* 447

Wing expanse: 32 mm

Diagnostic Characters: Male, upper forewing without band; spots yellow. Upper hindwing spots in 4 and 5 cojoined to form a single large spot; smaller spots in 2 and 3. Above, bases and much of hindwing clothed with ochreous hairs. Upper forewing spot in 2 large and quadrate.

Distribution: Reported from Meghalaya, Sikkim, Assam, West Bengal and northwest Himalaya. Elsewhere: Myanmar.

Habits and Habitat: Found upto 1,900 m in the Himalayas. A very uncommon species, occur where its food plant is grown and is a fast flier.

Genus: *Borbo* Evans

1949. *Borbo* Evans, *A. Catalogue of Hisperiidae from Europe, Asia, Australia in Brit. Mus. Nat. Hist.* 437: 44, 436

49. *Borbo bevani* (Moore, 1878) (Bevan's Swift)

1878. *Borbo bevani* Moore. *Proc. Zool. Soc. London*, 688

1949. *Borbo bevani* Evans, *A Catalogue of Hesperiidae from Europe, Asia and Australia in Brit. Mus. Nat. Hist.* 437

Wing expanse: 30 mm

Diagnostic Characters: Upper forewing with white spots, spot in 3 nearer to spot 2 than to 4; often with spot in 5. Under hindwing with small spots in 2 to 6, but 4 and 5 may be absent and some may appear upperwing. Female, upper forewing with spot in 1b; often present in male also.

A *Papilio protenor*

B *Papilio polyctor polyctor*

C *Pieris canidia indica*

D *Catopsilia crocale*

E ***Gonepteryx rhamni nepalensis*** **Male**

F ***G. r. nepalensis*** **female**

G ***Eurema hecabe fimbriata***

H ***Colias electo fieldi***

Plate 5.11: Some butterflies recorded in Khajjiar lake area

A *Danaus genutia*

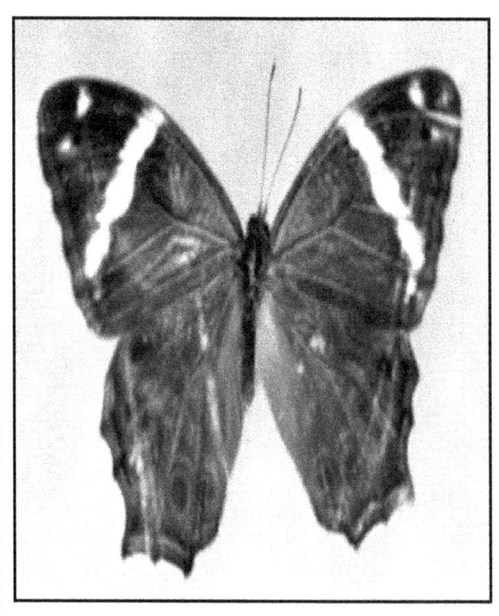

B *Lethe verma verma*

C *Lasiommata schakra schakra*

D *Aulocera saraswati saraswati*

E *Ypthima nareda nareda*

F *Ypthima ceylonica hubneri*

G *Ypthima sakra nikaea*

H *Melanitis leda ismene*

Plate 5.12: Some butterflies recorded in Khajjiar lake area

A *Athyma opalina*

B *Neptis hylas astola*

C *Pseudergolis wedah*

D *Cynthia cardui*

E *Childrena children*

F *Issoria lathonia*

G *Libythea myrrha*

H *Pseudozizeeria maha*

Plate 5.13: Some butterflies recorded in Khajjiar lake area

Distribution: Reported from Meghalaya, Assam, Sikkim and from Kashmir to Kumaon. Elsewhere, found in Bhutan, Indo-China, Malayan sub-region, Myanmar and Thailand.

Habits and Habitat: Found upto 2,500 m in the Himalayas; not abundant but had been found settling on garden flowers and *Lantana* in the hills. Present in cultivated areas in surroundings of Khajjiar.

G) ORDER: DIPTERA

Readily distinguished from other insects due to the presence of a single pair of wings; hind wings reduced to small knobbed structures called halters, which function as an organ of equilibrium. Majority of the species minute to small soft bodied insects with highly mobile head, large compound eyes, variable antennae and suctorial type mouth parts.

There is presence of 5 species of dipterans belonging to 5 genera, spread over 3 families from Khajjiar area of Chamba. Of these, *Machimus* is identified upto genus level and other species reported are *Chrysomya rufifacies, Calliphora vomitoria, Phaenicia Sericata* and *Musca domestica.*

Family: Asilidae

Commonly called Robberflies. Adults predaceous and attack a number of insects. Top of the head hollowed out between the eyes. Face more or less bearded. Thorax stout and legs strong and stout.

Genus: *Machimus* loew

1. *Machimus* sp.

(Plate 13 A)

Diagnostic Characters: Body stout and elongated; abdomen tapering; body thickly haired. Colour of the body, dark brown to black with yellow stripes; eyes black. A dense moustache of bristles on the face and 3 simple eyes in a characteristic depression between their two large compound eyes. Antennae short and three segmented. Thorax stout and bears strong legs.

Machimus is a very common species in Khajjiar area probably due to abundance of food in the form of a number of other small insects.

Family: Calliphoridae

Cosmopolitan insects of utmost economic importance. Most of the species of the size of housefly or little larger. Body shinning and metallic coloured often with blue, green, or black thoraxes and abdomen. Antennae 3-segmented; arista of antennae plumose at the tip. Frontal structure present, and also well-developed calyptras. A continuous dorsal suture across the middle, along with well-defined posterior calli, present at the thorax.

Genus: *Chrysomya* Robinean-Deasvoidy

1830. *Chrysomya* Robinean- Deasvoidy, *Myodaries*, p. 444

Eyes in males completely holoptic or separate, sometimes widely so. In females, eyes always widely separated. One pair of outwardly directed fronto-orbitals present. Short and fine bristles present on dorsum. Abdomen broad and oval. In males hypopygium sometimes strongly developed but usually small and inconspicuous.

2. *Chrysomya rufifacies* (Macquart, 1842)

(Plate 13 B)

1842. *Lucilia orientales* Macquart, *Mem. Soc. Roy. Sci. Arts Lille, Annee*, p. 302

Size: Length 10 mm

Diagnostic Characters: Eyes separate in males by a distance equal to width of third antennal segment. Antennae brown; prothorax and thorax greenish blue with purple reflection. Prothoracic spiracles white, a prostigmatic bristle present. First segment of thorax black in males and greenish in females. Second and third segments black banded on posterior margins. Wings transparent but infuscicated slightly at base of subcoastal cell. Pro, meso and metathoracic legs black in colour.

Distribution: Commonly distributed in Oriental and Australian regions. In India found in Chandigarh, Haryana, Punjab, Rajasthan, Jammu and Kashmir, Uttarakhand and Himachal Pradesh. Very commonly in Khajjiar area.

Genus: *Calliphora* Robineau-Desvoidy

1830. *Calliphora* Robineau-Desvoidy, *Myodaries*, p. 433

Eyes more or less approximated in male, widely separated in female. Abdomen has marginal bristles only on third and discal, and marginal bristles only on fourth visible segments. Lower squama of wing bears long upstanding hairs on at least part of the upper surface.

3. *Calliphora vomitoria* (Linnaeus, 1758)

(Plate 13 C)

1758. *Musca vomitoria* Linnaeus, *Syst. Nat.*, ed. x, i: 595

Size: Length 12 mm

Diagnostic Characters: Antennae dark brown, base of third segment red in colour. Parafrontalia dark grey, silver flecked below ocellar triangle and opposite point of insertion of antennae. Thorax dull bluish-black but slight silver dusted anteriorly. Abdomen shining blue to purple with very sparse silver dusting. Legs black. Wings hyaline with slight infuscated base, basicostal scale black, subcostal sclerite possess tawny pubescence. Halteres orange coloured.

Distribution: Reported from north America, Europe, China and India. In India found in Sikkim and western Himalayas.

Habits and Habitat: Found in diverse habitat and breeds in many kinds of decaying matter. Generally observed in Khajjiar meadow mainly during spring and summer seasons.

Genus: *Phaenicia* Robineau-Desvoidy

1863. *Phaenicia* Robineau-Desvoidy, *Posth. Ii*, xxx, p. 750

Metallic coloured flies of medium size. Parafrontalia and parafacilia covered by silver or golden tomentum. Wings hyaline. Legs brown to black. Middle tibia usually with one, sometimes with two or three antero-dorsal bristles.

4. *Phaenicia sericata* Meigen, 1826

(Plate 13 D)

1826. *Musca sericata* Meigen, *Sitz. Beschr.* v: 53

Size: Length 10 to 14 mm.

Diagnostic Characters: Parafrontalia covered with fine hairs. Thorax brilliant metallic blue-green or golden coloured with black markings and possesses three cross grooves and black bristles. Wings transparent with light brown veins. Abdomen shining green, evenly covered on dorsum and venter with short bristles. Antennae and legs black. Middle tibia with one antero- dorsal bristle.

Distribution: Cosmopolitan.

Habits and Habitat: Commonly associated with sheep but not host specific and plays significant role in forensic entomology. Commonly found in Khajjiar area.

Family: Muscidae

Well represented throughout the world. Generally members of this family have 3-segmented Antennae. Frontal suture present. Clypters well developed. Arista usually plumose throughout the length. Hypopleuron without bristles.

Genus: *Musca* Linnaeus

1758. *Musca* Linnaeus, *Harr. Exp.*, p. 32

5. *Musca domestica* Linnaeus, 1758

(Plate 13 E)

1758. *Musca domestica* Linnaeus, *Harr. Exp.*, p. 32

Size: Length 8-12 mm

Diagnostic Characters: Females slightly larger than the males. Eyes reddish and mouthparts sponging type. Space between the eyes narrow in males while females have a much larger space between their eyes. Thorax gray and bears four longitudinal dark lines on the back. The whole body covered with hair-like projections. Abdomen gray or yellowish with dark midline and irregular dark markings on the sides. The underside of the male yellowish.

Distribution: They are cosmopolitan and distributed throughout the world.

Habits and Habitat: Feed on faeces, open sores, sputum and moist, decaying organic matter, such as spoiled food, eggs, fruit and flesh. Take food in liquid form. Adults usually live 15 to 25 days, but may live up to two months. Survive longer in cooler temperatures. Common species in human settlements around Khajjiar meadow, not so common during winter months.

A *Machimus* sp.

B *Chrysomya rufifacies*

C *Calliphora vomitoria*

D *Phaenicia sericata*

E *Musca domestica*

Plate 5.14: Some dipteraus recorded in Khajjiar lake area

H) Order: Hymenoptera

Winged members have four membranous wings. The hindwings smaller than the forewings and have a row of tiny hooks on their anterior margin by which the hindwing attaches to the forewing. Wings contain relatively few veins, and in some minute forms have almost no veins. Mouthparts mandibulate, but in higher forms the labium and maxillae form a tongue like structure through which liquid food is taken. Antennae long and usually contain ten or more segments. The tarsi usually have 5 segments. The ovipositor well developed, but in some higher forms modified into a sting, which acts as a defence organ; only females can sting.

A total of 7 species of hymenoptera belonging to 6 genera and 5 families are recorded from Khajjiar area of Himachal Pradesh. These includes some species of economic importance viz., *Bombus trifasciatus, Apis dorsata, Apis cerana indica, Polistes hebraeus* and *Vespa basalis.*

Family: Apidae

This family includes bumblebees, honeybees and eulossinebees. Maxillary palps vestigial, genal area broad and corbiculae on the hind legs. Hind tibiae without apical spur and if present then jugal lobe of hindwing absent. Lack pygidial plate.

Genus: *Melecta* Latreille

1805. *Melecta* Latreille, *Hist. Nat. Ins.* xiv, p. 48

Head transverse and narrower than the thorax. Antennae very long and filiform. Basal joint of flagellum clavate. Labial palpi 4-jointed, basal joint thrice the length of the 2nd, apical two minute, while the maxillary palpi 5-jointed. Thorax subglobose, the scutellum bidentate. Forewing with the marginal cell narrow, rounded both at base and apex. Legs moderately long and pubescent. Abdomen conical at apex, truncate at base.

1. *Melecta himalayana* Bingham, 1897 (Cuckoo bee)

1897. *Melecta himalayana* Bingham, *The Fauna of British India, Hym.* vol. i. p. 516

Diagnostic Characters: Body densely pubescent; vertex of the head and base of the segments of the abdomen smooth. Thorax black. Clypeus, the face below the antennae and along the inner orbits of the eyes and the thorax covered with long silky pale yellow pubescence. A broad transverse stripe between the wings with pale brown pubescence present. Antennae and legs alutaceous, outside of the intermediate and posterior legs clothed with brown pubescence. Wings hyaline and slightly iridescent, with a brownish tint. Abdomen with the base of the 1st and 2nd segments sparsely, their margins and the apical segments densely, clothed with rich ferruginous pubescence.

Distribution: Present throughout the Himalayas.

Present in good numbers in Khajjiar area.

Genus: *Anthophora* Latreille

1804. Anthophora Latreille, *Hist. Nat. Ins.* xiv, p. 40

Head transverse less wide than the thorax. Antennae short and the scape shorter than the 2nd joint of the flagellum. Clypeus strongly produced and convex. Anterior margin of labrum rounded and nearly transverse. Labial palpi 4-jointed, the basal joint three times length of 2nd joint, apical two minute. Maxillary palpi 6-jointed, basal joint short, 2nd joint the longest. Forewing with the radial cell broad, rounded at apex, acute at the inner angle.

2. *Anthophora zonata* Linnaeus, 1758 (Flower bee)

(Plate 14 A)

1758. *Apis zonata* Linnaeus, *Syst. Nat.* ed. x. i, p. 576

Size: Length 11-13mm

Diagnostic Characters: Head and thorax densely pubescent; clypeus and the bases of the abdominal segments thinly pubescent; head and thorax finely and closely punctured under the pubescence. Front and vertex above antennae with dull rufo-fulvous pubescence. Side and apical margins of clypeus with a yellowish band; scape with yellowish spot; legs castaceous; femur of fore leg with white pubescence; tibia and tarsi with black bristles; middle tibia with one spine, hind leg with black hairs; apical margins of abdominal segments 1-4 with transverse bands of metallic blue scale-like hairs.

Distribution: Found throughout India, Myanmar, extending through the Malay regions to Australia.

Habits and Habitat: The species of this genus make their nests in the ground. Not very common species and can be observed a few times in Khajjiar area.

Genus: *Bombus* Latreille

1802. *Bombus* Latreille, *Hist. Nat Ins.*, p. 385

Densely pubescent; head not so wide in comparison to thorax, with front and face frequently elongate. Eyes narrow and do not reach down to the base of the mandibles. Antennae geniculate, long and filiform. Clypeus convex and much longer than broad. Labrum ciliated. Mandibles stout and broad, grooved at their apex. Labial palpi 4-jointed and maxillary palpi 2-jointed. Thorax globose and wings long. Legs stout and the posterior pair in the female smooth, with the tibiae and tarsi broad and flattened.

3. *Bombus trifasciatus* Smith, 1825 (Bumble bee)

(Plate 14 B)

1852. *Bombus trifasciatus* Smith, *Trans. Ent. Soc. N. S.* ii, p. 43

Size: Length, female 24-26 mm; male 19 mm

Diagnostic Characters: Head and face not elongate. Front and cheeks with a dense short pile, and on the front tufts of long pubescence. Vertex, space between eyes beneath, and base of mandibles bare, polished and shining. Thorax and abdomen densely pubescent. Anal segment on upper side, bare and finely punctured. A broad transverse band on the thorax. Legs, and the 3rd abdominal segment with black pubescence. Apical three abdominal segments with rich fulvous-red pubescence. Apex of the tibiae and the tarsi with ferruginous pile and wings fusco-hyaline.

Distribution: Present in India and China.

Habits and Habitat: Known to be an efficient pollinator mainly due to presence of fine hairs on the body and habit of feeding on pollen of flowers of a variety of crops. A common species in Khajjiar area, mainly around flowering areas.

Genus: *Apis* Linnaeus

1767. *Apis* Linnaeus, *Syst. Nat.* ed xii. i, p. 953

Head as wide as the thorax. Eyes ovate and pubescent. Antennae in a triangle shape on the vertex and clypeus elongate. Thorax globular. Forewings large and narrow. Radial cell long, narrow and rounded at apex; legs strong. The posterior tibiae smooth, shining and without spines. Margins of tibiae fringed with long hair. Females differ from the male in the head being narrower than the thorax. Posterior tibiae convex externally and without hairs. Abdomen much larger and sting straight.

4. *Apis dorsata* Fabricius, 1787 (Honeybee, Hive Bee)

(Plate 14 C)

1787. *Apis dorsata* Fabricius, *Ent. Syst.* ii, p. 328

Diagnostic Characters: Pubescence all over the head, thorax and abdomen. Head and mesonotum finely punctured under the pubescence. A short, medial, vertical groove present below the anterior ocellus. Head, thorax, legs, and apical three segments of the abdomen black. Pubescence fuscous on the head, front thorax, above legs and apical three segments of the abdomen. Pale ochraceous yellow on the hinder parts of the thorax and at the base of the abdomen and beneath legs. Wings pale fuscous or fusco-hyaline. Males much darker and eyes very large.

Distribution: Present throughout India, Burma and Ceylon. Also present in China, Malaya region and Java.

Habits and Habitat: Largest species of genus *Apis*. Builds its comb in wild exclusively, underside of brances of large plant, in caves or under over hanging rocks or in buildings. These shows social communities consisting of males, a single female called queen, and the workers. A few combs can be observed in Khajjiar area during summer months.

5. *Apis cerana indica* Fabricius, 1793 (Desi Makhi, Honeybee)

(Plate 14 E)

1793. *Apis cerana indica* Fabricius, *Ent. syst. Suppl.* p. 27

Diagnostic Characters: Head thorax and abdomen smooth and shining. Sparsely but sometimes densely pubescent. Head thorax and apical abdominal segment black. The scutellum and basal five segments of the abdomen testaceous yellow. Legs rufo-fuscous. Wings hyaline and iridescent. Males stouter in built, dark in colour and thorax densely pubescent. Females larger in size and darker in colour.

Distribution: One of the predominant species found in India, Pakistan, Nepal, Burma, Bangladesh, Sri Lanka, Thailand and mainland Asia.

Habits and Habitat: Commonly known as Indian Honeybee. Usually build multiple comb nests in some tree hallows and some manmade structures; also adapt to living in cavities in some human structures; less aggressive than other wild bees and also less swarming behavior and easily used for beekeeping. A common domesticated species in human settlements around Khajjiar area.

Family: Vespidae

Most of these insects black with yellow or whitish markings or brownish. Middle tibiae with 2 apical spurs; tarsal claws simple. Social wasps and the individuals in a colony of three castes i.e. queens, workers and males. Queens and workers have effective sting.

Genus: *Polistes* Latre.

1805. *Polistes* Latr. *Hist. Nat. Ins.* xiv, p. 348

Head flat in front and vertex not arched. Eyes small and reniform, not reaching the base of the mandibles. Antennae filiform, in the male arched at the apex. Clypeus pentagonal in shape and its anterior margin often roundly angular in the middle. Mandibles short and square in form, terminated by 4 teeth. Thorax elongate, the median segment oblique. Legs stout and moderately long; intermediate tibiae with two spines at apex, claws simple. Wings long.

6. *Polistes hebraeus* Fabricius, 1787 (Yellow wasp, Tamoori)

(Plate 14 E)

1787. *Polistes hebraeus* Fabricius *Syst. Piez.* p. 273

Size: Male 20-24 mm; female 13-16 mm.

Diagnostic Characters: Males and females same and yellow in appearance. Head, thorax and abdomen smooth, opaque, sometimes slightly shining. Mesonotum with two short, parallel, longitudinally impressed lines on its posterior portion. Medial groove on the median segment well-marked. Sutures on the head, thorax, and abdomen lined with black. A curved line on the vertex behind the ocelli present. Wings ferruginous or fulvo-hyaline. Abdominal segments with fine sinuate lines, a longitudinal brown line in middle of abdomen.

Distribution: Present throughout India, Burma and Egypt.

Habits and Habitat: Makes nests under the ground.

Genus: *Vespa* Linnaeus

1767. *Vespa* Linnaeus, *Syst. Nat.* ed. xii, i, p. 948

Head concave posteriorly and the cheeks much developed. Eyes variable in form, sometimes with, a considerable distance between them and the base of the mandibles, in a few cases extending down to the later. Antennae filiform and elongate in the male. Clypeus not terminated by a tooth. Mandibles very broad, terminated by 4 strong teeth. Thorax deep cubical but sometimes globose. Legs stout and short, the claws simple. Wings long.

A *Anthophora zonata*

B *Bombus trifasciatus*

C *Apis dorsata*

D *Apis cerana indica*

E *Polistes hebraeus*

F *Vespa basalis*

Plate 5.15: Some hymenopterans recorded in Khajjiar lake area

7. *Vespa basalis* Smith, 1852 (Social wasp)

(Plate 14 F)

1852. *Vespa basalis*, Smith, *Trans. Ent. Soc. New Ser.* ii, p. 4

Size: Female 20-23 mm; Male 17-19 mm

Diagnostic Characters: Head chestnut-red. Head, thorax, and abdomen smooth, with a fine silky lustre. Clypeus broader than long, deeply bi lobed anteriorly. Eyes large. Mesonotum has an anterior medial longitudinally impressed line. The basal segment with a short sub-apical transverse black band in the middle above, and some irregular black spots on each side near the base. Abdomen black. Wings flavo-hyaline, the costal and upper part of the medial cell in the forewing deep yellowish brown.

Distribution: Found in Sikkim, northern India and China.

Habits and Habitat: Make large nests of papery stuff, placed in trees, or in the ground, generally at the foot of a tree. Many species exceedingly irritable and fierce, resenting any intrusion, even at a good distance from their nests. Stings always painful, and very often dangerous. Seen predating on some honeybees in Khajjiar area.

III. MOLLUSCA

Soft-bodied animals, a majority of the members covered by a hard calcareous shell. The shell may consist of one, two or many pieces or sometimes may be internal and cartilaginous. Structurally heterogeneous group since a slug is strictly different in structure from a fresh water mussel or from an octopus or a snail.

During June-August only one species of a slug can abundantly seen in Khajjiar but it is not identified. A saddle-shaped mantle is present on upper side, behind the head, of the slug. Its mantle is positioned anteriorly and does not cover the entire length of the body. Two pairs of black and retractable tentacles are observed on its head. The species moves by rhythmic waves of muscular contraction on the underside of its foot by secreting a layer of mucus, which probably prevents any damage to the soft foot tissues. This species is seen mainly in moist environments in damp places (Plate 4 C and D).

B) VERTEBRATE FAUNA

A total of 100 species of vertebrates belonging to 83 genera, spread over 49 families and 22 orders are present in Khajjiar lake area (Table 5.3). Classwise analyses of data reveals that Class Aves dominates the fauna with 77 species belonging to 62 genera followed by Mammalia (16 species under 14 genera), Reptilia (4 species belonging to 4 genera), Amphibia (2 species spread over 2 genera) and Pisces (single species under a single genus) (Fig. 4). It is further seen that Khajjiar area supports a good population of some Vertebrate species like Carp (*Cyprinus carpio*), Himalayan Toad (*Bufo himalayanus*), Kashmir Rock Lizard (*Laudakia tuberculata*), Cattle Egret (*Bubulcus ibis*), Black Kite (*Milvus migrans*), Himalayan Griffon (*Gyps himalayensis*), Blue Rock Pigeon (*Columba livia*), Oriental Turtle-Dove (*Streptopelia orientalis*), Slaty-headed Parakeet (*Psittacula himalayana*), Common Cuckoo (*Cuculus canorus*), White-breasted

Kingfisher (*Halcyon smyrnensis*), Himalayan Pied Woodpecker (*Dendrocopos himalayensis*), Himalayan Bulbul (*Pycnonotus leucogenys*), Grey Bushchat (*Saxicola ferrea*), Streaked Laughingthrush (*Garrulax lineatus*), Rufous-bellied Niltava (*Niltava sundara*), Great Tit (*Parus major*), House Sparrow (*Passer domesticus*), Common Myna (*Acridotheres tristis*), Jungle Crow (*Corvus macrorhynchos*), Rhesus Monkey (*Macaca mulatta), Hanuman Langur (Semnopithecus ajax*), Himalayan Fox (*Vulpes vulpes*), Yellow throated Martin (*Martes flavigula*), Black Bear (*Ursus thibetanus*), Barking Deer (*Muntiacus muntjac*), Goral (*Nemarnhdus goral*), Flying Squirrel (*Petaurista petaurista*), House Mouse (*Mus musculus*) etc.

I. Pisces

A single species of Pisces i.e. *Cyprinus carpio* Linnaeus is present in the Khajjiar lake. It appears to be introduced in the lake as also revealed by the locals. Two different varieties of this fish i.e., *Cyprinus carpio communis* Linnaeus (Common Carp) and *Cyprinus carpio specularis* Lacepeds (Mirror Carp) are found in the lake.

Sub-Class: Actinopterygii

Order: Cypriniformes

Family: Cyprinidae

This family is commonly known as carp family. It is the largest family of fresh-water fish. Members of this family have oblique mouth which extends from the eyes. They have toothless jaws but pharyngeal teeth are present which helps in chewing. Snout is rounded or triangular in shape. Barbles are present on either side of mouth, and stout spine present on dorsal fin. Body have small and crowded scales.

Genus : *Cyprinus* Linnaeus

1758. *Cyprinus* Linnaeus, *Syst. Nat.* 10 ed.: 320

Snout is rounded or triangular in shape. Single long dorsal fin present, with 21 soft rays, and a stout saw-toothed spine in front of the dorsal and anal fins. Lateral line complete, with about 35 scales.

1. *Cyprinus carpio* Linnaeus, 1758 (Common Carp)

(Plate 15 A and B)

1758. *Cyprinus carpio* Linnaeus, *Syst. Nat.* 10 ed.: 320

Diagnostic Characters: Body laterally compressed, moderately elongate and covered with large cycloid scales. Snout blunt, mouth large and inferior. Two pairs of short barbells present on each side of upper jaw. Lips thick, mouth toothless, with the upper jaw slightly protruding. Lateral line complete, with about 35 scales. One long dorsal fin present with 21 soft rays, and a stout saw-toothed spine in front of the dorsal and anal fins. Pectoral fins with 14 rays. Pelvic fins thoracic of 8 rays, originating beneath origin of dorsal fin, one anal fin with 5 branched rays. Colour dark on back and golden on sides. Belly, pectorals and pelvics light-yellow, caudal fin gray with orange shade.

Distribution: Found in the rivers and streams, many natural lakes and impoundments, and some farm ponds, throughout India. Elsewhere distributed throughout the continental United States and extends from central Canada to central Mexico. Carp can tolerate a variety of environmental conditions and habitat types which has allowed them to invade such a large geographical area. The natural range of the species is composed of two regions, water bodies of the Ponto-Caspian-Aral basin and basins of the Far East rivers and rivers of south-eastern Asia, from the Amur River in the north to Yunnan (South China) and Burma in the south (Bogutskaya, 1998).

Habits and Habitat: Carp prefers water bodies with stagnant and slowly flowing waters. It feeds at temperature above 8-10^{0}C, reproduces, as a rule, at temperature above 15^{0}C. Feeding is heterotrophic, consumes zooplankton, vegetable and animal detritus, zoobenthos, macrophytes. The food spectrum changes with age depending on food supply in a water body. Young fish feed on plankton, overgrowing, and colonial green algae. Adults feed more on molluscs and aquatic plants.

Of the two varieties of carp, Mirror Carp, though very similar to Common can be distinguished from the later, by the presence of a irregular, patchy scale pattern, robust body with flakes in the dorsal and lateral part, and large scales resembling mirrors.

II. AMPHIBIA

Only two species of Amphibia belonging to two different families are present in Khajjiar pasture. A good population of *Bufo himalayanus* is recorded from the lake area especially during monsoon season, however.

Order: Anura

Family: Bufonidae

Animals of this family are known as true toads and are characterized by squat, plump bodies with short legs, and skin rough-warty. True toads are widespread and occur natively on every continent except Australia.

Genus : *Bufo* Laurenti, 1868

1708. *Bufo* Laurenti, *Syn. Kept.*, pp. 25

They have large parotid glands. They also lack an anterior breastbone and do not have teeth. Males develop dark nuptial pads on the thumbs and inner fingers that assist in amplexus. In most species, breeding males develop a dark throat.

1. *Bufo himalayanus* Gunther, 1864 (Himalayan Toad)

(Plate 15 C)

1864. *Bufo melanostictus* var. *himalayanus* Gunther, *Reptiles of British India*, pp. 422

1882. *Bufo himalayanus* Boulenger, *Fauna of British India*, pp. 505

Diagnostic Characters: One of the largest *Bufo* of India. Crown deeply concave, with low, blunt supra-orbital ridges; Snout short and blunt; inter-orbital space broad.

than the upper eyelid, tympanum very small and indistinct. Fingers free, first finger does not extend beyond second. Toes half to two-third webbed, with single subarticular tubercles; inner and outer metatarsal tubercles present; without tarsal fold. Dorsal side of the body with irregular, distinctly porous warts; parotids very prominent, large, elongate, at least as long as the head. Live colour brown. Males without vocal sacs.

Distribution: In India found in Sikkim, Meghalaya, Arunachal Pradesh, West Bengal, Uttar Pradesh, Jammu & Kashmir and Himachal Pradesh. Also present in Nepal.

Habits and Habitat: The species show defensive behaviour and when handled, secretes a corrosive fluid both from the parotids and warts on the dorsum. Pesticides and fertilizers used in the nearby orchards have adverse effects on this species. Also contamination of breeding grounds is a major threat to this animal. Present in good number from Khajjiar meadow area.

Family: Ranidae

Members of this family have smooth, moist-skin, with large, powerful legs and extensively webbed feet with pointed toes. Key characteristics of true frogs include bony breast bones and horizontal eyes. They also have slim waist and have toothed upper jaws. They vary greatly in size. Breeding for true frogs typically occurs during the spring with groups of males calling-in females to the breeding areas. Breeding males have swollen forelimbs and thumbs that facilitate pectoral amplexus. They are known as true frogs.

Genus : *Rana* Linnaeus 1700

1700. *Rana* Linnaeus, *Syst. Nat.* I: 354

The tibio-tarsal articulation reaches the tip of the snout, or beyond. On dorsal side skin smooth, seldom warty. Colour brown above and a black line on the canthus rostralis and on the temporal region.

2. *Rana liebigii* Gunther, 1830

(Plate 15 D)

1830. *Rana liebigii,* Gunther *P. Z.S.*, 157

1830. *Rana liebigii* Boulenger *Cat. Batr. Sal.*, 21

Diagnostic Characters: Head moderate and much depressed, snout very short and rounded. Inter orbital space nearly as broad as the upper eyelid and tympanum small and hidden. Fingers moderate, first finger not extending beyond the second. Toes also moderate, truncated or slightly swollen at the end. Toes entirely webbed and don't have any tarsal fold. Nuptial excrescences well developed; inner metatarsal tubercle oval, not very prominent; no outer tubercle. Lateral folds black-margined, legs indistinctly cross-barred. Male with internal vocal sacs. During the breeding season remarkable on account of the extreme thickness of the arms and of the patches of spinose warts on the breast, the inner side of the arms, and the inner fingers.

Distribution: In India, found in Sikkim, and in states of western Himalayas.

Habits and Habitat: Lives in a damp climate, but not so essentially aquatic in its habits but is found in damp jungles.

Emarks: *Rana liebigii* also known as *Rana vicina* (synonym) but is smaller in size than the later and found in western Himalayas.

III. REPTILIA

There are 4 species of reptiles belonging to 4 genera spread over 3 families and 1 order in Khajjiar area. It is recorded that family Colubridae is represented by a maximum of 2 species and other families namely Agamidae and Scincidae are represented by a single species each.

Order: Squamata

Family: Agamidae

Members of this family have scaly bodies, well-developed legs, and a moderately long tail. They cannot shed their tails and regenerate. Many agamid species are capable of limited change of their colours to regulate their body temperature. The key distinguishing feature of the agamids is their teeth, which are borne on the outer rim of the mouth. In some species, males are more brightly coloured than females and colours play a part in signalling and reproductive behaviours.

Genus: *Laudakia* Gray, 1845

1845. *Laudakia* Gray, *Cat. Liz. Brit. Mus.* p. 254

Throat and chest brownish, profusely spotted with dark blue; upper side of head light brown.

1. *Laudakia tuberculata* (Hardwicke & Gray, 1827) (Kashmir Rock Agama)

(Plate 15 E)

1827. *Agma tuberculata Hardwicke & Gray, Zool. Journ.*3: 218

1935. *Laudakia tuberculata Smith, Fauna Brit. Ind.* **2**: 214

Diagnostic Characters: Overall body sturdy with a flat head. Head depressed and elongated, tympanum large and distinct. Colour of the dorsal side dark-olive brown with numerous dark-brown spots on either side of a lighter vertebral line. Adult male specimens have bluish tinge on the dorsal side. Upper side of head light brown. Throat and chest brownish, profusely spotted with dark blue. Belly of adult as well as young ones, whitish. An elongated patch of enlarged scales present on the belly. Limbs moderately strong and toes longer. Fifth toe extends beyond the first toe. Tail depressed, longer than the head and body. Upper portion of the tail strongly keeled and almost equal scaled.

Distribution: Found in western Himalayas (Kashmir, Northern Punjab, Himachal Pradesh and Uttarakhand). In Himachal Pradesh, found throughout the state including the Trans-Himalayan districts of Lahoul & Spiti, and Kinnaur. Elsewhere: Afghanistan, Pakistan and Nepal.

Habits and Habitat: The species is diurnal and terrestrial, inhabiting the holes, crevices and such other rocky structures. Omnivorous in feeding habits and its food mainly comprises the insects like ants, small orthopterans, lepidopterns etc. Breeding season varies from May to August, lays 7-20 eggs in a single clutch. Frequent fighting breaks out between males.

Special comments: Some male adults can be observed to have beautiful and brilliant shades of bright yellow, orange bluish black, purple and black on shoulders, breast, flanks, under parts and throat during the months of May to August. These can be easily seen on rocky area near entrance gate of the Khajjiar wildlife sanctuary.

Family: Scincidae

This family is the largest of the sixteen or so families of lizards and members are generally called skinks. They look roughly like true lizards, but most species have no pronounced neck. They have relatively long-snouted and somewhat flattened skulls. The head is usually covered with enlarged plates. Some genera, have reduced limbs, lack forelegs, and in some genera number of digits less than five. In such species, locomotion resembles that of snakes more than that of lizards. Most species of skinks have long, tapering tails that can be shed during pursuit of life.

Genus: *Scincella* Mittleman, 1950

1950. *Scincella* Mittleman, *Herpetologica* 6 (2): 17-20.

2. *Scincella himalayanus Gunther, 1864* (Himalayan Ground Skink)

(Plate 15 F)

1864. ***Scincella himalayanus*** Günther, *The Reptiles of British India,* 452 pp.

Diagnostic Characters: A small skink having a bronze dorsum with indistinct lighter and darker markings and dark brown vertebral stripe. Lateral stripe of brass colour and having irregular margins. A broad, dorsal, dark-brown stripe emerges from snout and reaches up to the proximal part of tail through eye and upper side of the forelimbs. Another, lower, broad stripe bordered below by a narrow, irregular, white stripe edged with black; distal body colour bronzy, with numerous light and dark-brown spots. Top of the head and upperside of the limbs bronzy, with dark dots all over; belly bluish-white.

Snout small and bluntly pointed; ear-openings oval, smaller than eyes; the lower eyelids with undivided transparent disc. Limbs short and digits long and sub-cylindrical. Tail one and half times as long as the head and body.

Distribution: Found in Kashmir, Himachal Pradesh and Uttar Pradesh. Elsewhere: present in Pakistan, Nepal and southern Turkistan.

Habits and Habitat: Prefers damp areas or open grasslands. Also present in lake sides, banks of rivers and gardens. The species is insectivorous and viviparous (produces 3 or 4 young ones at a time). A good number of individuals of this species can be seen in meadow area around the lake.

Family: Colubridae

Members of this family are distinguished from other snakes primarily by the dentition, which usually comprises solid teeth on the maxilla, palatine, pterygoid and dentary, but never on the premaxilla. A few species have enlarged and/or grooved posterior maxillary teeth, which channel venom from the supralabial. Members have enlarged ventral scales in a single row, a more or less cylindrical tail in which all sub-caudals are divided.

Genus: *Amphiesma* Dumeril, Bibron & Dumeril, 1854

1854. *Amphiesma* Dumeril, Bibron & Dumeril, *Erpetologie generale ou Histoire Naturelle complete des Reptiles.* Vol. 7, 780 S.

3. *Amphiesma platyceps* (Blyth, 1854) (Eastern Keelback)

1854. *Tropidonotus platyceps* Blyth, *J. Asiatic Soc. Bengal.* 23: 297

Diagnostic Characters: Colourations variable, generally olive brown above, with small black spots; sometimes with a dorso-lateral series of white spots; frequently two white black-edged parallel lines, or an elliptical mark on the nape, or a white black-edged streak on each side of the head, or a black line near eye; lips white or yellow, belly yellowish, with or without blackish dots, bordered with bright red in living specimens; frequently a black line or a series of elongated black spots along each side of belly; lower surface of tail mottled with black; throat sometimes black.

Distribution: Found from western to eastern Himalayas in suitable habitat types.

Habits and Habitat: Prefers hill streams. This species can be observed at night in the pine forests near streams in the meadow area.

Genus: *Ptyas* Fitzinger, 1843

1843. *Ptyas* Fitzinger, *Systema Reptilium. Amblyglossae*, pp 106.

4. *Ptyas mucosus* (Linnaeus, 1758) (Indian Rat Snake)

1758. *Coluber mucosus* Linnaeus, *Mus. Ad. Frid*, 37, pl. 23

1864. *Ptyas mucosus* Gunther, *Reptiles of British India*, pp. 249

Diagnostic Characters: A large slender snake, reaches up to 3500 mm in length (Whitaker and Captain, 2004). Head distinctly broader than the neck. Live colouration olive green to brown, yellowish or greyish above, with irregular and strongly marked black cross-bars on the posterior half of the body; yellowish-white below, the posterior ventrals and sub-caudals edged with black; lips and throat white, the scales edged with black.

Distribution: Found throughout India. From Himachal Pradesh reported from all parts of the state except Trans-Himalayas. Elsewhere, Sri Lanka, Pakistan, Afghanistan, Nepal, Bangladesh, Myanmar and southern China.

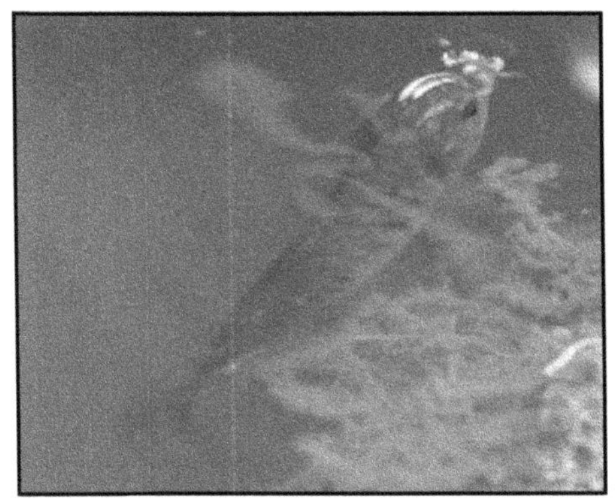

A *Cyprinus carpio specularis*

B *Cyprinus carpio communis*

C *Bufo himalayanus*

D *Rana liebigii*

E *Laudakia tuberculata*

F *Scincella himalayanus*

Plate 5.16: Some fishes, amphibians and reptiles recorded in Khajjiar lake area

Habits and Habitat: Feeds mainly on small mammals, toads and birds. Common in status, conservation status WLPA, 1972 schedule-II. Very few numbers can are seen in khajjiar, generally seen in summer.

IV. AVES

Birds dominate the khajjiar with 77 species belonging to 62 genera, 12 orders and 31 families (Table 5.1). In this area, Muscicapidae is the most represented family with 22 species belonging to 15 genera followed by Accipitridae and Corvidae (6 species each), and Paridae, Phasianidae, Columbidae and Picidae (3 species each). The study reveals that Khajjiar lake supports 20 such species of birds which are resident and rest 57 are seasonal-local and long range migrants. Of the 57 species, 35 are seasonal-local migrants, 4 are winter visitors and 10 are summer visitors. Moreover, Khajjiar lake supports 8 such species which shows winter and summer influx. Of these, 6 species shows summer influx, whereas, winter influx is shown by 2 species only (Table5.1, Fig. 5).

Analyses of data on relative abundance shows that 25 species of birds are very common, 30 are common, 21 are uncommon and 1 is rarely seen in the area (Fig. 5). Further analysis of residential status and relative abundance reveals that of the 20 resident species 10 are very common, 8 are common and 2 are uncommon of the 57 seasonal-local migrants, 15 species are very common, 24 are common, and 17 are uncommon and only 1 species is rare. Categorization of 4 winter visitors reveals that 2 are uncommon, 1 each is common and rare. Moreover, analysis of data on relative abundance of summer migrants shows that of the 10 species, 1 species is very common, 5 are common and 4 are uncommon. Grouping of the species among winter and summer influx reveals that of the 2 species which shows winter influx, 1 each is common and uncommon , whereas, out of 6 species that shows summer influx 1 is very common and 5 are common in the Khajjiar lake area

Analysis of feeding habits of these birds show that 36 species are insectivorous, 12 omnivorous, 9 frugivorous, 8 graminivorous. 4 species of these birds eats on aquatic animals, 4 scavengers, 3 carnivorous and only 1 species eats pure vegetable matter (Fig. 6).

Order: Ciconiiformes

Family: Ardeidae (Herons and Egrets)

Long-legged, lanky wading birds, with long, slender, flexible necks (in most species), a kink in the middle enables the neck to be retracted into a flat shaped 'S' in flight. Tarsi very long; toes long and slender, the middle and outer toes united by a small web at their base. Claw of middle toe comb like.

Genus: *Bubulcus* Bonaparte

1855. *Bubulcus* Bonaparte, *Compt. Rend. Acad. Sci. Paris*, 40: 722

Plumage pure white; feathers of head and neck orange-buff during breeding season.

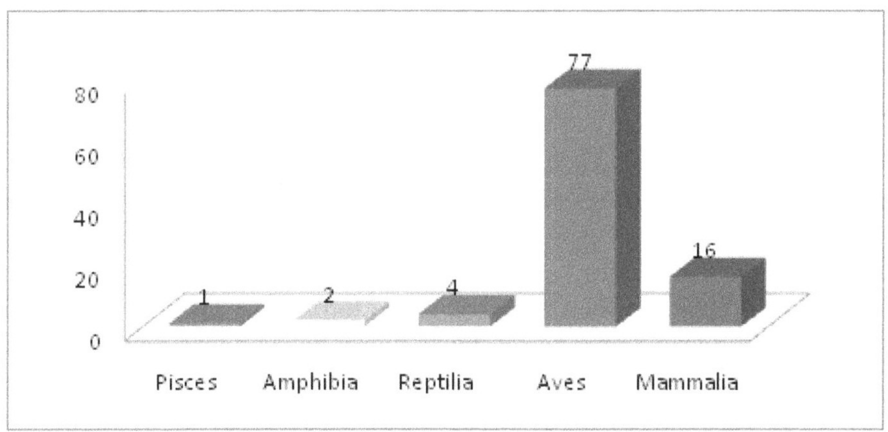

Figure.4: Number of species of vertebrate groups in Khajjiar area

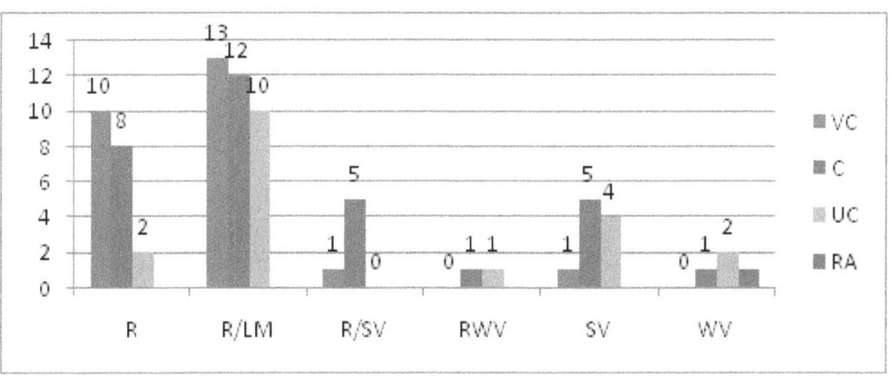

Figure. 5: Comparative residential status and relative abundance of birds of Khajjiar area

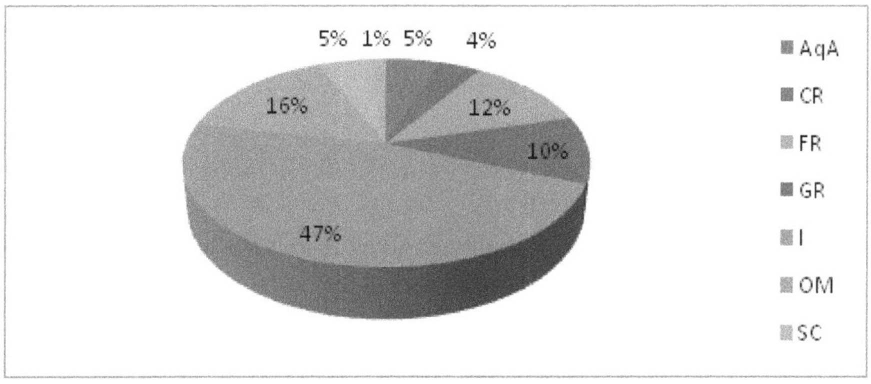

Figure. 6: Feeding habits of the birds in Khajjiar lake area

1. *Bubulcus ibis* (Linnaeus, 1758) (Cattle Egret)

(Plate 16 A)

1758. *Area ibis* Linnaeus, *Syst. Nat.*, ed. 10, 1: 144

Size: Hen =; 48-53 cm.

Diagnostic Characters: In non-breeding plumage, a lanky snow-white bird very similar to Little Egret. Always identified from Little Egret by yellow bill *contra* black. In breeding plumage, unmistakable golden buff on head, neck and back, feathers disintegrated and hair-like.

Status and Distribution: Resident, found throughout India up to 1500 m in Himalaya, migrating to lower elevations in winter.

Habits and Habitat: Gregarious while feeding and roosting. Usually seen in attendance on grazing village cattle on damp grassy margins of tanks as well as dry fallows and forest glades far from water. This bird is a summer migrant to the Khajjiar area.

Breeding: Breeds in mixed colonies almost round the year, though mainly from June to August in North India.

Order: Falconiformes

Family: Accipitridae (Hawks & Vultures)

Bill short with upper mandible longer than lower, curved and strongly hooked at tip; basal portion covered with a cere and nostrils brightly coloured. Feet strong, with powerful hooked claws. Hallux always present. Female usually larger.

Key to the Genera

A. Head and neck bare..1

B. Head and neck feathered..2

1. Bill slender ....a

   Bill stout...b

   a. Nostrils elongate and horizontal.....................*Neophron*

   b. Nostrils a narrow vertical slit..............................*Gyps*

   Nostrils round or oval.....................................*Aegypius*

2. Bill lengthened and vulturine in shape; claws blunt; a tuft of long bristles on chin ....*Gypeatus*

   Bill short and aquiline in shape; claws sharp and generally greatly curved; no tuft on chin c

   c. Tarsus completely feathered in front and behind...*Ictineatus*

   Tarsus naked or partially feathered.........................*Milvus*

Genus: ***Milvus*** **Lacepede**

1799. *Milvus* Lacepede, *Tabl. Ois.*: 4

2. *Milvus migrans* (Boddaert, 1783) (Black Kite)

1783. *Falco migrans* Boddaert, *Tabl Pl. enlum.*: 28

Size: Kite; 55-68.5 cm

Diagnostic Characters: Dark rufous-brown hawk with whitish crescent at the base of primaries on under-wing and deeply forked tail, both conspicuous in flight. Head brown, streaked with pale coloured feathers. A white patch present under the eye.

Status and Distribution: Found throughout India, resident in Himachal Pradesh showing some seasonal local movements; found up to 4500 m altitude (occasionally reaching 5300 m).

Habits and Habitat: Usually seen around towns, villages, mainly around slaughter-houses, fish markets, garbage dumping grounds and bazaars. Roosts in selected trees, in large numbers, during late evening hours. Largely omnivorous, feeds chiefly on offal and garbage in urbanized areas. Feeds on young chickens and duckling in poultry runs. Sometimes pick earthworms from ground. This is a resident and very common species in Khajjiar area.

Breeding: Season varies from March to May.

Genus: ***Gypaetus*** **Storr**

1784. *Gypaetus* Storr, *Alpenreise*: 69

3. *Gypaetus barbatus* (Linnaeus, 1758) (Bearded Vulture)

1758. *Falco barbatus* Linnaeus, *Syst. Nat.*, ed. 10, 1: 87

Size: Himalayan Griffon +; 122 cm.

Diagnostic Characters: A very big sized vulture with a wedge-shaped tail; feathers present all over the head and neck unlike most of the other vultures. Adult has blackish upperparts, streaked with white; rusty-white head and neck; rusty-white underparts contrasting with black under-wing coverts. A tuft of black bristle like feathers 'the beard' pendant under the chin. Juvenile has dark brown body with almost black head and grey-brown underparts.

Status and Distribution: Resident in Himalayas between 1200 to 4000 m, also observed by the Everest expeditions at 7500 m.

Habits and Habitat: Usually seen singly or in pairs, sailing around mountain slopes or scavenges around mountain villages. It also feeds on bones and marrow, which forms part of its diet. It carries the bone to a height of some 50 or 70 m and drops it on rocks below, which are often regular selected spots, from where it descends to the ground and swallows the pieces. Bearded Vulture is a resident and common species in Khajjiar area.

Breeding: Breeds from December to March.

Genus: ***Neophron*** **Savigny**

1809. *Neophron* Savigny, *Descr. Egypte, Ois.* 1: 68, 75, 76

4. *Neophron percnopterus* (Linnaeus, 1758) (Egyptian Vulture)

(Plate 16 B)

1758. *Vultur percnopterus* (*sic*) Linnaeus, *Syst. Nat.*, ed. 10, 1: 87

Size: Pariah Kite ±; 60-70 cm.

Diagnostic Characters: A small vulture with naked head, long and pointed wings, and wedge-shaped tail. Adult has almost entirely dirty-white plumage except black flight feathers and has naked yellow head, face and fore-neck. Juvenile has blackish brown body with bare grey face, head and fore-neck.

Status and Distribution: Resident almost throughout India except northeast India. It is partially altitudinal migrant to Himalayas.

Habits and Habitat: A very useful scavenger, keeps to the neighbourhood of human settlements, usually in groups of two or three. Feeds on carrion, offal and garbage and to a large extend on human ordure. Occasionally takes frogs and large crickets on grassland, also winged termites emerging from the ground. This species is a local migrant bird from Khajjiar area and analysed as a common species.

Breeding: Breeds from February to May.

Genus: ***Gyps*** **savigny**

1809. *Gyps* Savigny, *Descr. Egypte, Ois.*, 1: 68, 71

5. *Gyps himalayensis* Hume, 1869 (Himalayan Griffon)

(Plate 16 C)

1869. *Gyps himalayensis* Hume, *Rough Notes*: 12, 15

Size: White-backed Vulture +; 115-125 cm.

Field Diagnostics: Adult: buffish-white upper plumage with dark flight feathers and tail, and little-streaked pale brownish buff. Ruff, white-streaked pale brown. Legs and feet pinkish with dark claws. In overhead aspect, almost whitish underside contrasts with black trailing edges of wings and black tail. Juvenile: dark brown body and upper wing-coverts, boldly and prominently streaked with buff; back and rump also dark brown. Ruff brownish. Underparts dark brown, streaked with white.

Status and Distribution: Resident, found in Himalayas from western Pakistan to eastern Assam between 600 to 4000 m altitudes. Shows altitudinal movements in Himachal Pradesh.

Habits and Habitat: Found singly or in small groups around high mountain passes and also sails with outspread motionless wings over mountain tops and valleys. This is a resident and very common species in Khajjiar area.

Breeding: Breeds from January to April.

Genus: ***Aegypius*** **Savigny**

1809. *Aegypius* Savigny, *Descr. Egypte, Ois.*, 1: 63, 73

6. *Aegypius monachus* (Linnaeus, 1766) (Cinereous Vulture)

1766. *Vultur monachus* Linnaeus, *Syst. Nat.*, ed. 12, 1: 122

Size: Vulture +; 100-110 cm.

Diagnostic Characters: Very huge, uniformly black or blackish-brown Vulture with naked, pinkish neck surrounded by blackish ruff. Crown, ear coverts, lores and cheeks covered with black fur-like feathers. In flight, absence of white on crop and thighs distinguishes it from Red-headed Vulture. Juvenile blackish and more uniform than adult.

Status and Distribution: Breeds in Pakistan and Northwest Himalayas, in dry temperate zone between 1800 to 3600 m (Ali & Ripley, 1968-75). Winters mostly in Northwest and North India.

Habits and Habitat: A rare species often recorded feeding on carcasses alongwith other vultures. Prefers a wide range of habitats but mostly avoid forest. Feeds on carcases mainly, also recorded as hunting the tortoises and extracting the meads from under the carapace (Ali & Reply, 1983). This species is considered as a winter visitor in Khajjiar area.

Breeding: Breeds from March to April.

Genus: ***Ictinaetus*** **Blyth**

1843. *Ictinaetus* Blyth, *Jour. Asiat. Soc. Bengal* 12: 128

7. *Ictinaetus malayensis* (Temminck, 1822) (Black Eagle)

1822. *Falco malayensis* Temminck, *Planch. Color. d'Ois, livr.*: 20, pl. 117

Size: Kite+; 69-81 cm.

Diagnostic Characters: At rest, long wings extend to tip of the tail. Adult, dark brownish black with yellow cere and feet. In flight, long, broad and round-tipped wings raised in 'V' due to upturned primaries. Also shows whitish barring on upper tail-coverts, and faint greyish barring on tail and underside of wings. Juvenile has buffish head, underparts and under wing coverts, all streaked with dark.

Status and Distribution: Resident in Himalayas up to 2700 m and hills of India.

Habits and Habitat: Pairs can be seen sailing above the forest canopy. Eggs and nestlings of birds form an important part of its diet. Preferred habitat is evergreen and moist deciduous forests in hilly biotope. Feeds on large insects, frogs, laggards, rodents and other birds. Reported to kill jungle fowl and pheasants. Black Eagle is a local migrant and uncommon species in Khajjiar area.

Breeding: Season varies from January to April in the North, and November to March, in South India.

Order: Galliformes

Family: Phasianidae (Pheasants)

The 'Game Birds' which have formed a valuable food resource for man from earliest times come under this family. They are characterized by their terrestrial habits with stout unfeathered legs for progress on the ground. Claws short, blunt and very strong for scratching the ground for food, wings short and rounded, contour feathers with well-developed after-shaft.

Key to the genera

A. No ocelli on tail or tail-coverts..1

1. Wing over 200 mm..a

a. Tail longer than wing...I

Tail shorter than wing...II

I. Tail slightly longer than wings..........................b

b. Sides of head feathered................................*Pucrasia*

Sides of head unfeathered ....*Lophura*

II. Plumage of males metallic....................*Lophophorus*

Genus: *Pucrasia* G.R. Gray

1841. *Pucrasia* G.R. Gray, *List Gen. Bds.*, ed. 2: 79

8. *Pucrasia macrolopha* (Lesson, 1829) (Koklass Pheasant)

1829. *Satyra macrolopha* Lesson, *Dict. Sci. Nat.*, ed. Levrault, 59: 196

Size: Domestic fowl; Male: 58-64 cm, female: 52-56 cm.

Diagnostic Characters: Male has metallic-green head and ear-tufts, with a brown backwardly projected crest and metallic green horn like tuft of plumes on either side of the crest. A white patch on either side of upper neck. Overall body, brown streaked buff with some chestnut on breast and also on upperparts. Tail, chestnut brown and pointed. Female has streaked body with a small backwardly directed tuff and white throat.

Status and Distribution: Resident seasonal altitudinal migrant in Western Himalayas. Found in summer from 2000 to 4000 m and in winter from 1000 m to 2300 m.

Habits and Habitat: Mostly observed singly or in pairs, close in cover and difficult to flush, found in temperate and sub-alpine forests. Omnivorous in feeding habits, feeds on shoots, buds, tubers, leaves, berries acorns, seeds and insets. Recorded as a local migrant and common species from Khajjiar area.

Breeding: Breeds from May to June.

Genus: *Lophophorus* Temminck

1813. *Lophophorus* Temminck, *Pig. et Gall.* 2: 355

9. *Lophophorus impejanus* (Latham, 1790) (Impeyan Monal)

(Plate 16 D)

1790. *Phasianus impejanus* Latham, *Index Orn.* 2: 632

Size: Domestic hen +; 72 cm.

Diagnostic Characters: Male: shining green, copper and purple in colour, with a large white-patch on black and cinnamon tail, and with a crest of wire-like spatulate-tipped metallic-green feathers. Jet black on underside. Female: brownish with streaks of pale and dark brown, and has a short tuft on head; throat white; a crescent of white upper tail-covers, easily observed.

Status and Distribution: Resident, Himalaya between 2500 m to 5000 m altitude.

Habits and Habitat: Found mostly singly or in parties of 3 to 4 birds in rhododendron and deodar forest heaving open areas or pastures and moist hills sides with grass and weeds. Feeds on grass and flower seeds, roots, tubers, shoots, berries. Also feeds on insets and their larvae. This is an uncommon species in the area of Khajjiar.

Breeding: Breeds between April and June.

Genus: ***Lophura*** **Fleming**

1822. *Lophura* Fleming, *Philos. Zool.* 2: 230

10. *Lophura leucomelanos* (Latham, 1790) (Kaleej Pheasant)

(Plate 16 E)

1790. *Phasianus lencomelanos,* Latham, *Index Orn.* 2: 633

Size: Domestic fowl; Male 65-73 cm, female 50-60 cm.

Diagnostic Characters: Both sexes have naked red face and downsized tail. Male has blue-black upperparts, and variable amounts of white on rump and underparts, depending upon races. Male of *L. l. hamiltoni,* of Western Himalayas, has whitish crest. Female has brownish body, fringed with greyish-buff.

Status and Distribution: Resident, found in Himalayas and Northeast India, from foothills to 3700 m, but more common at lower altitude.

Habits and Habitat: Found in pairs or small family parties. Comes out to feed in the open in the early morning and afternoons, on forest track and freshly sown fields. Roost on trees in groups. Prefers forest with dense undergrowth. Feeds on grains, seeds, shoots, insects and their larvae, also small reptiles. A good population of this species is observed in Khajjiar areas.

Breeding: Breeds from March to July, varies with race and altitude.

Family: Recurvirostridae (Stilts)

Long-legged marsh birds; bill very long, straight.

Genus: ***Himantopus*** **Brisson**

1760. *Himantopus* Brisson, *Orn.* 1: 46; 5: 33

Nostrils slit like; wings long and pointed; tail short and square, sexes alike.

11. *Himantopus himantopus* (Linnaeus, 1758) (Black-winged Stilt)

1758. *Charadrius himantopus* Linnaeus, *Syst. Nat.*, ed. 10, 1: 151

Size: Grey Partridge ±; 35-40 cm.

Diagnostic Characters: A lanky black-and-white wader with fine straight bill; pointed black wings; long, thin pinkish legs which trail a long way behind tail in flight. Adult male has a few black spots on head; mantle and wings metallic black; pale grey-brown on tail; rest of the plumage white. Adult female has brownish-grey head and hind neck; black of male (mantle and wings) replaced by brown. Juvenile has browner upperparts with buff fringes.

Status and Distribution: Resident throughout India up to 1500 m in Himalaya. It migrates locally with water conditions.

Habits and Habitat: Usually found in small flocks, occasionally in groups of hundreds; feeds by wading in water or on dry mud, sometimes floats on water for feeding. Prefers freshwater and brackish marshes and saltpans. Feeds on molluscs, worms, aquatic insects and small seeds of sedges and marsh plants. An altitudinal migrant species considered as a summer visitor to Khajjiar sanctuary.

Breeding: Breeds between March and August.

Order: Columbiformes

Family: Columbidae (Pigeons and Doves)

Bill varied, short and slender; wings long form; tarsus short, more or less feathered; tail moderate, tail of twelve feathers (always of 14 feathers in *Treron*); tarsus elongated, feet more fitted for walking on the ground.

Key to the Genera

A. Plumage with much yellow-green above and below..............*Treron*

B. Tail of twelve feathers; plumage of upperparts or underparts, or both, not barred with black...1

1. Wing above 200 mm, or if 180-200 mm, outer tail-feathers blackish at tips..*Columba*

   Wing below 180 mm. or if 180-200 mm., outer tail-feathers white or grey at tips.......................................*Streptopelia*

Genus: ***Columba*** **Linnaeus**

1758. *Columba* Linnaeus, *Syst. Nat.*, ed. 10, 1: 162.

12. *Columba livia* Gmelin, 1789 (Blue Rock Pigeon)

(Plate 16 F)

1789. *Columba livia* Gmelin, *Syst. Nat.* 1 (2): 769

Size: Dove; 33 cm.

Diagnostic Characters: A bluish-grey pigeon, with metallic green, purple and magenta sheen on the neck and back; broad black bars across the wings. Tail grey with blackish terminal band.

Status and Distribution: Resident throughout India, except parts of Northwest and Northeast. It reaches 3300 m altitude in Himalayas.

Habits and Habitat: Lives in colonies, around rocky cliffs and gorges in the hills, old housings, factory sheds, residential buildings etc. Feeds on cereals, pulses, ground nuts, weed seeds, small tubers and green shoots of ground crops. Reported as a very common and resident species in Khajjiar.

Breeding: Breeding season varies mainly from April to August but also at other times, except in winter at higher altitudes.

Genus: *Streptopelia* Bonaparte

1855. *Streptopelia* Bonaparte, *Compt. Rend. Acad. Sci. Paris* 40: 17; id., Consp. Av. 2: 63.

13. *Streptopelia orientalis* (Latham, 1790) (Oriental Turtle-Dove)

1790. *Columba orientalis* Latham, *Index Orn.* 2: 606

Size: Pigeon; 33 cm

Diagnostic Characters: Adult has rufous-scaled scapulars and wing-coverts, brownish-grey to dusky maroon-pink underparts, grey lower back and rump, and a black-and-white chess-board mark on sides of neck. Tail black with grey sides and tip.

Status and Distribution: Resident and migrant (depending upon races), found in Himalayas, Northeast India, south to central peninsular India. Two of the races i.e. *S. o. meena* and *S. o. agricola* breeds during summer in Himalayas (up to 4000 m altitude) and migrate southwards in winter.

Habits and Habitat: Found in pairs or small groups, forming into larger flocks during migration. Feeds on scattered grains and weed seeds, in stubbles etc. in open forest. This is a very common and local migrant species in Khajjiar area.

Breeding: Season varies from May to July.

Genus: *Treron* Vieillot

1816. *Treron* Vieillot, *Analyse*: 49

14. *Treron sphenura* (Vigors, 1832) (Wedge-tailed Green-Pigeon)

(Plate 17 A)

1832 (1831). *Vignago sphenura* Vigors, *Proc. Zool. Soc. London*: 173

Size: Pigeon; 33 cm.

Diagnostic Characters: A yellowish-green pigeon with long wedge-shaped tail; indistinctly yellow-edged wing-coverts and tertials; dark-green rump and tail. Male has maroon patch on back and scapulars, and orange wash to crown and breast. Female like male but lacks orange wash to crown and breast, and maroon on upperparts.

Status and Distribution: Resident, local-seasonal and altitudinal migrant, found in Himalayas from plains level to 2500 m altitude. Summer visitor to hill stations and upland valleys.

Habits and Habitat: Usually found in small flocks on higher trees on leafy fruit-laden twigs. Prefers broad-leaved hill forests. Exclusively frugivorous, feed on drupes and barriers of tress. Present as a common species in Khajjiar area showing some summer influx in populations.

Breeding: Breeds mainly from April to June.

Order: Psittaciformes

Family: Psittacidae (Parakeets)

Arboreal fruit and grain eating birds usually of brightly coloured plumage, all the Indian representatives being chiefly green.

Genus: ***Psittacula*** **Cuvier**

1800. *Psittacula* Cuvier, *Lecons d'Anat. Comp.* 1

Bill short, stout, strongly hooked: upper mandible loosely articulated with the skull and capable of movement. Feet zygodactylus-2 toes in front, 2 behind. Primaries 10; tail-feathers 12.

15. *Psittacula himalayana* (Lesson, 1832) (Slaty-headed Parakeet)

1832. *Psittacula (Conurus) himalayanus* Lesson, *in. Belanger's Voy. Ind. Orient., Zool.*: 239

Size: Myna +; 41cm

Diagnostic Characters: Overall, grass-green parakeet, with dark-grey head bordered by a black stripe arising from black chin. Bill red-and-yellow; steeply graduated yellow-tipped tail. Male has a bright red shoulder patch, which is absent in female. Juvenile has entirely green plumage; dark head acquired after the first winter.

Status and Distribution: Resident, found in Himalayas from 600 to 2500m, occasionally down to 250m.

Habits and Habitat: Found in small to large flocks; keeps more to forests, also well-wooded areas. Feeds on nuts, acorns, seeds and fruits, wild as well as cultivated. Very destructive to walnuts, apples and pears in hill orchard. It is a local pest of maize. Observed as a local migrant and very common species in Khajjiar area of Himachal Pradesh.

Breeding: Season varies from March to May.

16. *Psittacula cyanocephala* (Linnaeus, 1766) (Plum-headed Parakeet) (Plate 17 B)

1766. *Psittacus cyanocephalus* Linnaeus, *Syst. Nat.*, ed. 12, 1: 141

Size: Myna -; 36 cm.

Diagnostic Characters: A yellowish or grass-green, slender parakeet, with plum-coloured head; a blackish collar; a maroon-reddish patch on the wing-shoulder. Narrow, blue central tail-feathers with broad white tips, conspicuous especially in flight. Head of female is duller and greyer, surrounded by a bright yellow collar. Ring on neck absent; maroon-patch absent or obsolete. Young, overall green; tips of primaries more pointed than in adults; central tail-feathers shorter and broader.

Status and Distribution: Resident and local migrant almost throughout India, except parts of Northwest and Northeast, found up to 1300 m altitude.

Habits and Habitat: Affects moist deciduous biotope; less closely associated with man-than Rose-ringed Parakeet. Mainly restricted to well wooded country, light forests and cultivation. Feeds on all kind of grains and fruits, buds, nectar of flowers. Also destructive to orchard fruits. Recorded as a summer influx and common species in Khajjiar area.

Breeding: Breeds from January to April.

Order : Cuculiformes

Family: Cuculidae (Cuckoos)

First and fourth toes directed backwards. Contour feathers without any after-shaft, dorsal feather-tract divided between the shoulders and enclosing a lanceolate naked patch on the back; a nude oil gland present. Young nidiculous. Tail-feathers ten in all Oriental genera.

Genus: *Cuculus* Linnaeus

1758. *Cuculus* Linnaeus, *Syst. Nat.*, ed. 10, 1: 110

Tail rounded or graduated; plumage of upperparts not metallic. Wing 150 mm or more, long and pointed

17. *Cuculus canorus* Linnaeus, 1758 (Common Cuckoo)

1758. *Cuculus canorus* Linnaeus, *Syst. Nat.*, ed. 10, 1: 110

Size: Pigeon ±; 32-34 cm.

Diagnostic Characters: Adult male has dark ashy-grey upperparts; blackish brown tail, spotted and tipped with white (without the sub-terminal black band of Indian Cuckoo); underparts including chin, throat and breast pale-ashy; rest of the underparts white, narrowly barred with blackish. Adult female has rufous tinge on breast. Female has another phase with barred chestnut-and-blackish-brown upperparts; barred pale-chestnut and blackish throat and upper breast; rest of the underparts slightly rufous.

Status and Distribution: Breeds in Himalayas, Northern, Northeastern and Central India; scattered winter records from different places in India. Optimum zone for breeding 600 to 4100 m altitude, highest 5250 m in Kashmir.

Habits and Habitat: Mainly arboreal; usually seen singly, on leafy tree-tops or canopy of trees, occasionally descends to the ground. Prefers forest and well-wooded areas. Feeds on insects and their larvae, particularly hairy caterpillars. Common Cuckoo is a summer visitor of the Khajjiar area.

Breeding: Brood-parasite mainly on babblers, flycatchers, warbles, chats, pipits and shrikes.

Order : Strigiformes

Family: Strigidae (Owls)

Owls have large and rounded heads, big and forwardly facing eyes surrounded by broad facial disc, short neck and tail. Most of them are nocturnal and cryptically coloured.

Genus: *Glaucidium* Boie

1826. *Glaucidium* Boie, *Isis von Oken,* Bd. 2, col. 970

Middle claw not pectinate; ear tufts absent; wing 240 mm. and below; plumage barred above.

18. *Glaucidium cuculoides* (Vigors, 1831) (Asian Barred Owlet)

1831. *Noctua cuculoides* Vigors, *Proc. Zool. Soc. London*: 8

Size: Myna ±; 23 cm

Diagnostic Characters: A hornless, dark or olive-brown owlet, heavily barred with buff and whitish on upper as well as underparts; longitudinal dark striations on whitish abdomen and a prominent white throat patch. Juvenile has bars only on wings and tail; head and nape spotted with pale rufous; underparts streaked.

Status and Distribution: Resident and altitudinal migrant, Himalayas and Northeast India, from foothills up to 2700 m.

Habits and Habitat: Diurnal in habits. Generally found sitting out on bare branches or dead tree stumps during day light and freely feeding throughout the day. Found in tropical, subtropical and temperate forests. Feeds on beetles, grasshoppers, cicadas and other large insects, lizards, mice and small birds. This bird is a local migrant and common species in the area of Khajjiar.

Breeding: Season varies from April to May.

Order: Caprimulgiformes

Family: Caprimulgidae (Nightjars)

Bill short, weak, flexible and with an enormous gape. Nostrils tubular.

Genus: *Caprimulgus* Linnaeus

1758. *Caprimulgus* Linnaeus, *Syst. Nat.,* ed. 10, 1: 193

Wing long: 2nd primary (as.) generally longest.

19. *Caprimulgus macrurus* Horsfield, 1821 (Large-tailed Nightjar)

1821. *Caprimulgus macrurus* Horsfield, *Trans. Linn. Soc.* xiii (1): 142

Size: Pigeon ±; 33 cm.

Diagnostic Characters: A large sized brownish-buff nightjar, heavily marked with black and buff, with large head and a long broad tail. Crown and nape brownish-grey with bold black streaks down centre; pale rufous-brown nuchal collar; large white throat patch; scapulars, buff edged with black centres; coverts boldly tipped with buff. Male has white spots on four primaries and white tips to two outermost tail-feathers. Female lacks white primary spots or has smaller buff spots on primaries, and white tips to two outermost tail-feathers.

Status and Distribution: Resident, found from Himalaya south to Central India, up to 1800 m altitude.

Habits and Habitat: Crepuscular and nocturnal in habits. Found usually singly or pairs; keeps well camouflaged amongst dry leaves in day time. Also perches freely on trees. Noticed by resonant calls, *Chaunk-chaunk-chaunk.* Prefers edges of tropical and subtropical forests. Feeds upon moths, beetles and other dusk and night flying insects. Large-tailed Nightjar is a summer visitor and common species in Khajjiar area.

Breeding: Breeds mainly from March to May.

Order: Apodiformes

Family: Apodidae (Swifts)

Bill very small, much hooked; wings excessively long and pointed; tail usually short, of 10 feathers only; hind-toe directed inward but reversible to the front.

GENUS: *Apus* Scopoli

1777. *Apus* Scopoli, *Intr. Hist. Nat.*: 483.

Shafts of tail-feathers not spiny; anterior toes with three phalanges each. All four toes directed forward, though first reversible.

20. *Apus affinis* (J.E. Gray, 1830) (House Swift)

1830. *Cypselus affinis* J.E. Gray, *in* Gray and Hardwicke's *Ill. Ind. Zool.* 1(2),

pl. 35, f. 2

Size: Sparrow; 15 cm.

Diagnostic Characters: Overall, smoky-black with long-narrow wings, short-square or slight fork-tail, conspicuous white rump and throat.

Status and Distribution: Resident throughout India up to 2000 m altitude; locally migratory in coldest months.

Habits and Habitat: Usually seen flying in small parties around habitations, cliffs and ruins. Spends most of the day on wings. Feeds on dipterans, tiny flying bugs and beetles, winged ants and air bone spiders. This bird is a resident and common species in Khajjiar area of Himachal Pradesh.

Breeding: Breeds practically all round the year, except the coldest months i.e. November to February.

Order: Coraciiformes

Family: Alcedinidae (Kingfishers)

Small to medium-sized birds with long strong bill; large head; short neck and legs; short and rounded wings. Flight is strong wing rapid wing beats and most of them are usually found near water systems.

Key to the Genera

A. Plumage black and white.......................................*Megaceryle*

- Plumage not black and white..................................*Halcyon*

Genus: *Halcyon* Swainson

1821 (1820-21). *Halcyon* Swainson, *Zool. Ill.* 1, text to pl. 27

21. *Halcyon smyrnensis* (Linnaeus, 1758) (White-breasted Kingfisher)

(Plate 17 C)

1758. *Alcedo smyrnensis* Linnaeus, *Syst. Nat.*, ed. 10, 1: 116

Size: Myna +; 28 cm.

Diagnostic Characters: A shining blue kingfisher with chocolate-brown head, neck and underparts. Long, heavy and pointed red bill. Throat and centre of breast shining white. A large white wing patch conspicuous in flight. Colour of upperparts slightly varies seasonally.

Status and Distribution: Resident with some seasonal local migrations, found throughout India, up to an altitude of 2300 m.

Habits and Habitat: Mostly found singly or in pairs around wet paddy fields, village tanks, canals, wells, streams, gardens etc. Feeds largely on insects, like grass hoppers, crickets, scarab and other beetles. Also feeds on small animals sometimes on fish. A local migrant from Khajjiar area.

Breeding: Season varies mainly between April and July.

Genus: *Megaceryle* Kaup

1848. *Megaceryle* Kaup, *Verh. Nat. Vereins Hessen*, ii: 68

22. *Megaceryle lugubris* (Temminck, 1834) (Greater Pied Kingfisher)

1834. *Alcedo lugubris* Temminck, *Planch. Color. d'Ois., livr.*: 92, pl. 548

Size: House Crow ±; 41-43 cm.

Diagnostic Characters: A crested, black-and-white kingfisher, with an erectile spotted-and-streaked white-and-black crest; boldly spotted and barred with white on upperparts and tail. A white half collar on nape; a band of black spots on breast mixed with rufous; sides of abdomen, flanks and under tail-coverts barred.

Status and Distribution: Resident, found in Himalaya and Northeast India, from terai and foothills up to 2000 m.

Habits and Habitat: Found in pairs around the mountain rivers, large rivers in foothills. Raises the crest and switches-up the tail over the back from time to time. Feeds mainly on fish. A resident species of Khajjiar area is common in abundance.

Breeding: Breeding varies from March to April.

Family: Upupidae (Hoopoes)

Bill long and slender, slightly curved throughout; the tip acute and entire; nostrils small.

Genus: *Upupa* Linnaeus

1758. *Upupa* Linnaeus, *Syst. Nat.*, ed. 10, 1: 117

Wings rounded; tail moderate to long; tarsi short and stout; toes and claws strong.

23. *Upupa epops* Linnaeus, 1758 (Common Hoopoe)

(Plate 17 D)

1758. *Upupa epops* Linnaeus, *Syst. Nat.*, ed. 10, 1: 117

Size: Myna; 31 cm.

Diagnostic Characters: An orange-buff bird with black-and-white zebra markings on back, wings and tail, and a black-tipped crest (fan shaped when erected).

Status and Distribution: Summer visitor to far north up to 4600 m altitude; resident and winter visitor to almost rest of India.

Habits and Habitat: Found singly or in pairs and in small parties of 8 to 25 birds when migrating. Erects the crest and expands the tail from time-to-time. Prefers open country, cultivation and villages. Entirely insectivorous, feeds largely upon underground grubs and pupae of beetles, crickets, earwigs, locusts and grass hoppers, ants and termites. This is a common and winter visitor species in Khajjiar.

Breeding: Breeds from April to June.

Order: Piciformes

Family: Capitonidae (Barbets)

Bill stout and somewhat conic, inflated at the sides, moderate in length or short, wide at the base, compressed towards the tip; base of upper mandible with stiff bristles projecting forwards.

Genus: *Megalaima* G.R. Gray

1842. *Megalaima* G.R. Gray, *Appendix to List Gen. Bds.*: 12

Culmen generally blunt; wings and tail short, with 10 tail feathers; toes in pairs, the hind-claws much curved.

24. *Megalaima virens* (Boddaert, 1783) (Great Barbet)

(Plate 17 E)

1783. *Bucco virens* Boddaert, *Tabl. Pl. Enlum.*: 53

Size: Myna +; 33 cm

Diagnostic Characters: It has large yellow bill; blue-black head; olive brown breast and mantle; green wings and tail; olive streaked-yellowish underparts and red under tail-coverts.

Status and Distribution: Resident with seasonal-altitudinal movements, in Himalayas and Northeast India. Found from 1000 to 3000 m in summers and descends down to foothills during winters.

Habits and Habitat: Found singly or in small groups of five to six birds, mainly in subtropical and temperate forests. Feeds chiefly on fruits-drupes and berries. Also feeds insects like beetles, cicadas, flying termites, hornets etc. Great Barbet is considered as a local migrant species in Khajjiar area.

Breeding: Season varies from April to July.

Family: Picidae (Woodpeckers)

Bill moderate and long, straight, angular, wedge-like; tongue long, extensible; wings moderate, or rather long; tail of twelve feathers, ten of them with shafts, thick and stiff, the outermost pair minute; feet with the toes in pairs, one toe sometimes wanting.

Key to the Genera

A. Mantle wholly or partly green, back not barred......................*Picus*

- Mantle black and white..*Dendrocopos*

Genus: ***Dendrocopos*** **Koch**

1816. *Dendrocopos* Koch, *Syst. baierischen Zool.* 1: xxvii, 72, pl. 1 A

25. *Dendrocopos himalayensis* (Jardine and Selby, 1831) (Himalayan Pied Woodpecker)

1836. *Picus himalayensis* Jerdine & Selby, *in* Jardine's *Ill. Orn.*, 3, Sig. D., pl. 116

Size: Myna ±; 23-25 cm.

Diagnostic Characters: Crown and crest crimson; black rear border to ear coverts; upperparts black with a broad white shoulder patch; small white spots and barrings on wing-quills; underparts fulvous and unstreaked; under tail-coverts crimson. Female has black on crown instead of crimson.

Status and Distribution: Resident, Western Himalayas between 1500 m to treeline.

Habits and Habitat: Found in pairs, working around tree-trunks and on moss-covered branches in forests. Consumes insects chiefly grubs of beetles collected from under the bark or moss, largely also seeds of various confers. This is a local migrant and very common species in Khajjiar area.

Breeding: Breeding season varies from April to June.

Genus: *Picus* Linnaeus

1758. *Picus* Linnaeus, *Syst. Nat.*, ed. 10, 1: 112

26. *Picus squamatus* Vigors, 1831 (Scaly-bellied Green Woodpecker)

1831 (1830-31). *Picus squamatus* Vigors, *Proc. Zool. Soc. London* (1): 8

Size: Pigeon +; 35 cm

Diagnostic Characters: A green woodpecker, with dull grass-green upperparts; white streaks on outer edges of wings; yellow rump; white streaking on tail. Underparts greyish-green with black scaling on lower breast and vent; white supercilium; white cheek stripe; a black moustachial stripe. Male has crimson crown and crest. Female has blackish crown and crest finely spotted with white.

Status and Distribution: Resident in Himalayas, found between 1000 to 3300 m.

Habits and Habitat: Usually seen in pairs, working its way up in a series of jerky spurts, also descends to ground. Found in forest, scrub, and open areas with large trees. Feeds mainly on ants and termites, also on larvae and pupae of wood boring beetles. Also feeds on berries, especially in winters. This bird is a resident but is uncommon species in Khajjiar.

Breeding: Season extends from March to June.

27. *Picus canus* Gmelin, 1788 (Black-naped Green Woodpecker)

1788. *Picus canus* Gmelin, *Syst. Nat.* 1 (1): 434

Size: Pigeon ±; 32 cm

Diagnostic Characters: Upperparts yellowish-green, with white bars on dark brown wing quills; yellow rump; indistinctly barred tail; underparts uniform greyish-green; face plain-grey. Male has crimson crown and has black nape and moustachial stripe whereas, female has black crown.

Status and Distribution: Resident in Himalayas, Northeast and Eastern India, from terai and foothills to 2000m altitude.

Habits and Habitat: Typically like woodpeckers, found in forests. Feeds mainly on ants and termites, also larvae and pupae of wood boring beetles, also feeds on barriers in winters, flower nectar as supplement food in season. Observed as a resident species from Khajjiar area.

Breeding: Season varies from May to June.

Order: Passeriformes

Family: Hirundinidae (Swallows)

Slender and slim bodied birds with long and pointed wings; small bill; weak legs; more or less forked tail. Long-tailed ones are Swallows, whereas, shorter-tailed ones are Martins. They hawk day-flying insects in swift and sustained flight.

Genus: *Hirundo* Linnaeus

1758. *Hirundo* Linnaeus, *Syst. Nat.*, ed. 10, 1: 191

Tarsus and toes bare, or with a tuft of feathers only. Upper plumage glossy blue-black; tail with or without spots, or unglossed brown with tail-spots.

28. *Hirundo daurica* Linnaeus, 1771 (Red-rumped Swallow)

1771. *Hirundo daurica* Linnaeus, *Mantissa Plant*: 528

Size: Sparrow ±; 16-17 cm.

Diagnostic Characters: Upperparts, glistening blue-black, with rufous-orange neck sides; rufous-orange rump; underparts finely streaked buffish-white; tail deeply forked. Six races in the subcontinent, differing mainly in strength of streaking on underparts and rump.

Status and Distribution: Summer visitor to Northwest India, reaching up to 3300 m in Himalayas; resident and winter migrant to rest of India.

Habits and Habitat: Found in pairs or in small flocks, hawking insects in company with swifts and other swallows. In summers found in upland cultivation and grassy hills, winters in open country and forest clearings. Feeds on gnats, midges and tiny flying beetles and bugs, winged termites and ants etc. Reported as a common species in Khajjiar area with summer influx.

Breeding: Season varies from April to August.

Family: Motacillidae (Wagtails)

Small and slim ground-feeding birds with slender pointed bill; pointed wings; fairly long legs. Wagtails are generally unstreaked, brighter, wag their tail up and down, often call in flight and roost communally in non-breeding season.

Genus: *Motacilla* Linnaeus

1758. *Motacilla* Linnaeus, *Syst. Nat.*, ed. 10, 1: 184

Upper plumage unstreaked; tail comparatively long.

29. *Motacilla maderaspatensis* Gmelin, 1789 (Large Pied Wagtail)

1789. *Motacilla maderaspatensis* Gmelin, *Syst. Nat.* 1: 961

Size: Bulbul =; 21 cm.

Diagnostic Characters: Male has white head and back, with a prominent white supercilium and large white wing-bar. Outer rectrices white. Chin, throat and breast black, and rest of the underparts white. Female similar to male but might be having black of chin and breast, replaced by grey.

Status and Distribution: Resident, shows some seasonal local movements. Found throughout India up to 2200 m in Himalayas.

Habits and Habitat: Feeds on small beetles, locusts, dragon flies, snakes and small seeds, along water edges; vigorously wags its tail up and down; keeps in pairs. Often perches on housetops. This species is a local migrant and but uncommon species in Khajjiar area.

Breeding: Season varies from March to June in the North and December to June in South India.

Family: Campephagidae (Minivets)

Bill of moderate length, or rather short, rather deep vertically, broadish at base; culmen arched or curved; rectal bristles few; nostrils basal; wings of moderate length; third and fourth, or fourth and fifth quills sub-equal and longest; tail long, rounded, or graduated; feathers of back and rump often rigid; tarsus short; feet weak and moderate.

Genus: *Pericrocotus* Boie

1826. *Pericrocotus* Boie, *Isis von Oken,* col. 972

Tail more or less graduated. Shafts of rump-feathers spiny. Tail strongly graduated, outer rectrices less than half length of the tail.

30. *Pericrocotus ethologus* Bangs and Phillips, 1914 (Long-tailed Minivet)

(Plate 17 F)

1914. *Pericrocotus brevirostris flavillaceus* (*sic*) Bangs and Phillips, *Bull. Mus. Comp. Zool.,* Harvard Univ., 58: 283

Size: Bulbul; 20 cm

Diagnostic Characters: Male has glossy, blue black upperparts, including head, neck, back and throat; rump scarlet; red wing line down secondaries; outer tail feathers crimson; underparts deep crimson. Female has indistinct yellow forehead and supercilium; grey ear-coverts and pale yellow throat.

Status and Distribution: Breeds in Himalaya and northeast India, winters south to Central India.

Habits and Habitat: Found in flocks; arboreal in habits; found in forests. Insectivorous in feeding habits. This bird is a summer visitor and very common species in Khajjiar area.

Breeding: Breeding season varies from April to July.

Family: Pycnonotidae (Bulbuls)

Bill generally short, straight and depressed; rictal bristles well developed; nostrils exposed; wings moderate or long; legs and feet very short, only suited for perching.

Key to the Genera

A. Bill equal to, usually shorter than tarsus. Difference between longest primary and longest secondary less than length of tarsus ....*Pycnonotus*

- Bill longer than tarsus. Difference between longest primary and longest secondary more than length of tarsus ....*Hypsipetes*

Genus: *Pycnonotus* Boie

1826. *Pycnonotus* Boie, *Isis von Oken*, col. 973

31. *Pycnonotus leucogenys* (Gray, 1835) (Himalayan Bulbul)

(Plate 18 A)

1835. *Brachypus leucogenys* Gray, *in* Hardwicke's *Ill. Zool.* 2, pl. 35, fig. 3

Size: Red-vented Bulbul ±; 20 cm.

Diagnostic Characters: An earth-brown coloured bird, with a forwardly-curving, pointed-crest; white patches on cheeks and black throat. Under tail-coverts yellow and tip of tail white.

Status and Distribution: Resident, Himalayas between 300 to 2400 m altitude.

Habits and Habitat: Found in pairs or in small-flocks in open scrub jungles and hillsides having wild berries and around cultivations, towns and villages. Feeds on fruits and berries. Nectar, seeds and insects also taken. In status a resident and very common species in Khajjiar area.

Breeding: Mainly breeds between March and June.

Genus: *Hypsipetes* Vigors

1831. *Hypsipetes* Vigors, *Proc. Zool. Soc. London*: 43

32. *Hypsipetes leucocephalus* (P.L.S. Muller, 1776) (Black Bulbul)

(Plate 18 B)

1776 (=Gmelin, 1788). *Turdus leucocephalus* P.L.S. Muller, *Syst. Nat.* 1 (2): 829

Size: Red-vented Bulbul ±; 23 cm.

Diagnostic Characters: A black-crested, slaty-grey bulbul with slightly forked tail, and red bill, legs and feet. Juvenile without crest, has brown bill and feet, and whitish breast and belly.

Status and Distribution: Resident with local altitudinal movements; summer visitor to higher parts. Breeds between 1000 to 2400 m, and descends to foothills during winters.

Habits and Habitat: Mainly found in pairs or small parties and prefers broad leaved forest and plantations. Feeds mainly on fruits and berries, also on insects and nectar. This bird is considered as common and resident species.

Breeding: Season starts from end of April and ends by September.

Family: Laniidae (Shrikes)

Bill short or of moderate length, notched or toothed at the tip; gape rather wide with rictal bristles.

Genus: *Lanius* Linnaeus

1758. *Lanius* Linnaeus, *Syst. Nat.*, ed. 10, 1: 93

Tarsus short, strong, usually with large scutae in front and on the toes.

33. *Lanius vittatus* Valenciennes, 1826 (Bay-backed Shrike)

(Plate 18 C)

1826. *Lanius vittatus* Valenciennes, *Dict. Sci. Nat.* (Levrault) 40: 227

Size: Bulbul; 17 cm.

Diagnostic Characters: Adult has black band across forehead to ear coverts, contrasting with pale grey crown and nape; chestnut maroon back; a white patch at base of primaries; white rump. White edged black tail; underparts largely white, rusty on flanks. Juvenile has uniform greyish upperparts; lacks white wing patch and has rufous tail.

Status and Distribution: Resident throughout India, except parts of Northeast. Shows seasonal movements mainly in Northern India, reaches up to 2000 m in Himalayas during summer.

Habits and Habitat: Usually seen solitary or in pairs, inhabiting the same area day after day, perching on some exposed perch, around open dry scrub, and bushes in cultivation. Feeds on any small living creature that can be over powered by this bird like insects, lizards, frogs, fledgling, sickly or disabled birds, young fields mice and earthworms. This bird is a winter visitor and uncommon species from Khajjiar area.

Breeding: Season varies from April to July in North India.

Family: Muscicapidae

Bill rather wide, depressed, shallow; the culmen straight, distinctly hooked and notched at the tip; rictal bristles numerous and strong; wings moderate; tail generally short or moderate; tarsus short, weak; feet moderately small, feeble.

Key to the Genera

Juvenile plumage spotted ....A

Juvenile plumage unspotted ....B

A. Bill slender and fairly strong..1

Bill broad and flat..2

1. a. Tail shorter than wing; tarsus long. Sexes dissimilar...*Monticola*

b. Bill stout, shorter than head laterally compressed, hooked at tip. Wing rounded....................................*Myiophonus*

c. Bill about half the length of head. Rictal bristles moderate. Well demarcated patch not clearly visible on underwing..*Turdus*

d. Bill short and slender; rictal bristles short. Wing pointed

Tail shorter. Rictal bristles well develop...........*Rhyacornis*

Tail rounded. Wing long but with large first primary equal to half the length of second. Tarsus long and strong..*Chaimarrornis*

e. Bill strong and fairly stout. Wing long. Tail long, graduated and deeply forked. Lower mandible convex.........*Enicurus*

f. Bill rather less than half the length of head, notched, broad at base. Rictal bristles strong. Wing pointed...........*Saxicola*

2. Rictal bristles short...*Muscicapa*

Rictal bristles well developed.......................................g

g. Throat and breast slaty with rufous gorget ....*Ficedula*

Upper plumage mostly blue.......................................h

h. Entire throat blackish blue ....*Niltava*

Throat entirely blue ....*Eumyias*

B. Bill small, slender; legs short and weak................................3

Bill very variable; legs and feet strong 4

3. Bill shorter than head, slim and pointed; rectal bristles present but short..*Phylloscopus*

Bill broad and blunt, about half the length of the head. Rictal bristles greatly developed. Wings rather rounded; second and third primary graduated....................................*Seicercus*

4. Bill strong and straight. Legs strong. Wing short and rounded, often edged with a brighter colour.......................*Garrulax*

Bill shorter than head, slender and curved; nostrils covered by a membrane, rectal bristles moderate. Tail long and graduated...*Heterophasia*

Genus: ***Monticola*** **Boie**

1822. *Monticola* Boie, *Isis von Oken* 1: 552

34. *Monticola cinclorhynchus* (Vigors, 1832) (Blue-headed Rock-Thrush)

1832. *Petrocincla cinclorhyncha* Vigors, *Proc. Zool. Soc. London*. 172

Size: Bulbul; 17 cm

Diagnostic Characters: Adult male (summer) has blue head and throat; a blue shoulder patch and a white wing-patch; a broad band on head side and back; tail blue black; rump and underparts (except blue throat) orange. Winter and non-breeding male has pale fringes to black back and blue head. Female has olive-brown upperparts, spotted with brown; underparts white.

Status and Distribution: Summer visitor to Himalayas and Northeast India, between 1200 to 2200m and winters mainly in Western Ghats and Assam hills; uncommon and rare elsewhere in central India.

Habits and Habitat: A forest bird largely arboreal (occasionally descends to ground), found singly or in pairs, found in open dry forests during summers. Feeds chiefly on insects, also other small animals such as frogs and lizards. Occasionally, seeds, barriers and flowers nectar also taken. This species have a status of summer visitor and uncommon species in Khajjiar area.

Breeding: Season extends from May to July.

35. *Monticola rufiventris* (Jardine and Selby, 1833) (Chestnut-bellied Rock-Thrush)

(Plate 18 D)

1833. *Petrocincla rufiventris* Jardine & Selby, *Ill. Orn.* 3, pl. 129

Size: Myna ±; 24 cm

Diagnostic Characters: Male has blue upperparts including head, with some black on lores, ear-coverts and sides of neck. Throat, blackish-blue and rest of underparts chestnut-red. Female has buff patch on sides of neck; darkly-bared slaty olive-brown upperparts, and heavily scaled brown-buff underparts. Non-breeding male has buff fringed mantle, scapular and throat.

Status and Distribution: Resident and altitudinal migrant, Himalayas and Northeast India. Breeds between 1200 to 3300 m (optimum 1800 to 2400 m) and winters from 1800 m down to the foothills (occasionally plains).

Habits and Habitat: Found singly or in pairs, perching upright on bushes, trees etc; feeds mostly on ground, in open forest on rocky slopes. Feeds on beetles, tipulids, grasshoppers, butterflies, moths and other insects; also feeds on berries and seeds. This bird is a local migrant and uncommon species in Khajjiar area.

Breeding: Season varies from April to July.

Genus: ***Myiophonus*** **Temminck**

1822. *Myiophonus* Temminck, *Pl. col.* 2 (29), pl. 170

36. *Myiophonus caeruleus* (Scopoli, 1786) (Blue Whistling-Thrush)

(Plate 18 E)

1786. *Gracula caeruleus* Scopoli, *De. Flor. et. Faun. Insubr.* 2 (42): 88

Size: Pigeon; 33 cm.

Diagnostic Characters: A dark purple-blue bird, spangled with shining blue, which is brighter on forehead, shoulders, wing edges and tail. Yellow bill and a few white spots on wing coverts, both conspicuous.

Status and Distribution: Resident and altitudinal migrant, Himalayas and Northeast India. Breeds between 1500 to 2400 m (Occasionally 1000 m to treeline) and winters from 2400 m down to foothills.

Habits and Habitat: Seen singly or in pairs around small streams, running through jungles and cliffs, also seen around bungalows and road sides. Also observed away from water. Feeds mainly on earthworms, snails, crabs, larvae and aquatic insects but also take hatchling birds and almost all small creature. Also feeds on berries and some vegetable matter. This Thrush is a common species but is a local migrant species in Khajjiar area.

Breeding: Season varies from end of April to August. Two broods reared in a season.

Genus: *Turdus* Linnaeus

1758. *Turdus* Linnaeus, *Syst. Nat.*, ed. 10, 1: 168

37. *Turdus albocinctus* Royle, 1840 (White-collared Blackbird)

1840. *Turdus albocinctus* Royle, *Ill. Bot. Himalayan Mountains* 1 (1839): ixxvii, ixxviii

Size: Myna +; 27cm

Diagnostic Characters: Male, entirely black, with a broad white collar and throat. Under tail-coverts with white streaks. Female, rufous-brown with variable pale collar. Juvenile has buff underparts with dark-drown spottings.

Status and Distribution: Resident-altitudinal migrant in Himalayas from Chamba eastwards to northeast India. Breeds between 2100 to 4200 m and winters from 3000 m down to hill bases (900 m).

Habitats and Habitat: Seasonally found solitary, in pairs or in groups in forest and forest edges. Feeds both on insects, and fruits and berries. Vegetable food probably predominant in autumn and winter. White-collared Blackbird is a local migrant and uncommon bird species in Khajjiar.

Breeding: Season varies from May to July.

38. *Turdus boulboul* (Latham, 1790) (Grey-winged Blackbird)

1790. *Lanius boulboul* Latham, *Index Orn.* 1: 80

Size: Myna +; 28 cm

Diagnostic Characters: Male is overall black, with a conspicuous greyish wing panel; yellow eye ring and orange bill. Belly and under tail-coverts with whitish fringes. Female, olive-brown, with pale rufous-brown wing panel. Juvenile has orange-brownish streakings on upperparts and variable orange-buff spottings on underparts.

Status and Distribution: Resident and altitudinal migrant, Himalayas and Northeast India. Found in summers (breeds) between 1800 to 2700 m and winters between 1200 to 2100 m, occasionally reaching foothills and adjacent plains.

Habits and Habitat: A shy bird found solitary, in pairs or in small flocks (depending upon season); feeds on ground, in forests and forest edges. Feeds on insects, earthworms, berries and fruits. Found to be a resident bird species with population influx during winter season in Khajjiar area.

Breeding: Season varies from March to August.

39. *Turdus viscivorus* Linnaeus, 1758 (Mistle Thrush)

1758. *Turdus viscivorus* Linnaeus, *Syst. Nat.*, ed. 10, 1: 168

Size: Myna +; 28 cm.

Diagnostic Characters: Upperparts, pale grey-brown, with white-edged flight feathers and white-tipped outer rectrices. Underparts buffish, boldly spotted with rounded dark-brown spots. Juvenile has whitish spottings on upperparts.

Status and Distribution: Resident, shows altitudinal movements in Western Himalayas. Breeds between 1800 to 3600m and descends down to 1200m, occasionally reaching foothills in winters.

Habitats and Habitat: Found singly, in pairs or in flocks, hopping on ground or on trees; in open coniferous forest, shrubberies (summer) or on grassy slopes and forest edges (winters). Feeds mainly on insects and their larvae, also on berries. This is a local migrant and uncommon species in Khajjiar area.

Breeding: Season varies from April to July.

Genus: *Chaimarrornis* Hodgson

1844. *Chaimarrornis* Hodgson, *in* Gray's *Zool. Misc.*: 82

40. *Chaimarrornis leucocephalus* (Vigors, 1831) (White-capped Redstart)

(Plate 18 F)

1831. *Phoenicura leucocephalus* Vigors, *Proc. Zool. Soc. London*: 35

Size: Bulbul; 19 cm.

Diagnostic Characters: A prominent white cap on black head; upper back and wings black; rump and tail rufous, with a broad black terminal tail band. Throat and breast black; rest of underparts chestnut.

Status and Distribution: Resident and altitudinal migrant, found in Himalayas and Northeast Indian hills. Breeds between 1800 to 5300 m and winters mostly below 1500 m down to the foothills.

Habits and Habitat: Keeps singly or in pairs, on boulders amidst streams and rivers, sometimes, especially in open country also found away from water. Food chiefly consists of insects but also consumes berries. Considered a common and local migrant species in Khajjiar area.

Breeding: Season varies from May to August.

Genus: *Rhyacornis* Blanford

1872. *Rhyacornis* Blanford, *Jour. Asiat. Soc. Bengal* 41: 51

41. *Rhyacornis fuliginosus* (Vigors, 1831) (Plumbeous Redstart)

1831. *Phoenicura fuliginosa* Vigors, *Proc. Zool. Soc. London*: 35

Size: Sparrow -; 12 cm.

Diagnostic Characters: Male, entirely slaty-blue, with rufous-chestnut tail and rufous lower belly. Female has dark grey-brown upperparts, with two rows of white spots on wings; white on base of tail and outer rectrices (partly). Underparts grey with white spottings.

Status and Distribution: Resident, found in Himalaya and shows altitudinal movements. Breeds between 1200 to 3900 m and winters from 2400 m down to the foothills.

Habits and Habitat: Found singly or in pairs, confined to mountain streams flitting from boulder to boulder. Continuously open and close its tail with a scissors-like action. Feeds chiefly on insects and occasionally consumes berries. A local migrant and common species of Khajjiar area.

Breeding: Season varies from April to July.

Genus: ***Enicurus*** **Temminck**

1822. *Enicurus* Temminck, *Pl. col.* (19), pl. 113

42. *Enicurus maculatus* Vigors, 1831 (Spotted Forktail)

(Plate 19 A)

1831. *Enicurus maculatus* Vigors, *Proc. Zool. Soc. London*: 9

Size: Bulbul; 25 cm.

Diagnostic Characters: A long-forked-tailed bird, with white forehead and fore-crown; a broad white wing bar and white rump; white spotting on mantle. Outer rectrices white and others black, tipped with white. Throat and breast black, whereas, rest of underparts white.

Status and Distribution: Resident and altitudinal migrant, found in Himalaya and Northeastern Indian hills. Breeds (summers) between 900 to 3600 m and winters from 2300 m down to 600 m.

Habits and Habitat: Mostly observed solitary or in pairs. Long forked-tail always raised horizontally, well off the ground, and gently swayed up and down; on stones in rocky streams in forest. Feeds on aquatic insects and small molluscs. This is a local migrant and uncommon species in Kalatop-Khajjiar areas.

Breeding: Season varies from April to July.

Genus: ***Saxicola*** **Bechstein**

1803. *Saxicola* Bechstein, *Orn. Taschenb.*, (1802), 1: 216

43. *Saxicola torquata* (Linnaeus, 1766) (Common Stonechat)

(Plate 19 B)

1766. *Muscicapa torquata* Linnaeus, *Syst. Nat.*, ed. 12, 1: 328

Size: Sparrow -; 13 cm.

Diagnostic Characters: Breeding male has black upperparts, having a large white patch on sides on neck and sides of breast; white wing patch and white on rump, and tail black. Throat black; breast orange, which changes to buff on belly. Non-breeding (winter) male has rufous-brown edged upperparts and rusty rump. Female has streaked rufous-brown upperparts; rufous breast and pale fulvous underparts.

Status and Distribution: Winter visitor, resident and altitudinal migrant (depending upon races) in Himalaya and Northeast Indian hills, breeds between 1500 to 3000m. Winters down to the foothills below 2200m south to almost entire India.

Habits and Habitat: Usually seen solitary or in pairs. In summers prefers open country with bushes, including high-altitude semi-desert, and winters in scrub, reed beds and cultivation. Food chiefly consists of ants and small beetles; also feeds on locusts and other insects. Common Stonechat is a local migrant and very common species.

Breeding: Breeding season extends from March to July.

44. *Saxicola ferrea* Gray, 1846 (Grey Bushchat)

(Plate 19 C)

1846. *Saxicola ferrea* Gray, *Cat. Mamms. Bds. Nepal*: 71, 153

Size: Sparrow; 15 cm.

Diagnostic Characters: Upperparts of male vary from ashy-grey to black, and has a white supercilium; black sides of head, and black tail with white outer edges. A white wing patch visible only during flight. Underparts light-grey, except white throat. Female has lightly-streaked brownish upperparts, with a buff supercilium, and rusty rump and tail sides. Throat white, rest of underparts light chestnut-brown.

Status and Distribution: Resident and altitudinal migrant, breeds in Himalayas and northeast Indian hills, between 1500 to 3300 m, and winters from around 2100 m down to foothills and plains.

Habits and Habitat: Usually found solitary or in pairs, perching on bush-tops, flittering its tail, in bushy area and secondary growth. Feeds on insects and on some seeds. Recorded as a local migrant and very common species in Khajjiar area.

Breeding: Season varies from April to July.

Genus: *Garrulax* Lesson

1831. *Garrulax* Lesson, *Traite d'Orn.*: 647

45. *Garrulax lineatus* (Vigors, 1831) (Streaked Laughingthrush)

(Plate 19 D)

1831. *Ciclosoma lineatus* Vigors, *Proc. Zool. Soc. London*: 56

Size: Bulbul ±; 20 cm.

Diagnostic Characters: A small brownish and greyish-brown bird, streaked with fine white (mantle and underparts), and dark brown (rest of the body). Rump and tail olive brown; tail rounded faintly barred, and with a greyish white terminal band. Ear coverts and wings rufous.

Status and Distribution: Resident, Himalayas, shows some seasonal movements. Found in summers between 1800 to 3000m (occasionally 1200 to 3900m) and in winters between 1000 to 1800m (occasionally 600 to 2750m).

Habits and Habitat: Keeps in pairs or in small parties, on the ground between low bushes, grass etc., in bush covered slopes, wooded nullahs, cultivation, under growth in open forests, gardens, etc. Feeds on insects, berries and seeds, occasionally even bread - crumbs. This bird is a very common and local migrant species in Khajjiar area.

Breeding: Season varies from March to September and at least two broods are reared.

46. *Garrulax variegatus* (Vigors, 1831) (Variegated Laughingthrush)

1831. *Cinclosoma variegatus* Vigors, *Proc. Zool. Soc. London*: 56

Size: Myna ±; 24 cm

Diagnostic Characters: Grey-olive in overall appearence, with black patches on silvery-greyish wings; rufous-buff forehead and malar stripe; black throat. Crown, nape and ear-coverts dark grey with a short white streak behind eye. Back, rump and upper tail-coverts olive-brown. Tail black, distally grey with white tip.

Status and Distribution: Resident in Western and Central Himalayas, shows some vertical movements. Found in summers between 2400 m to treeline (occasionally lower range is 1800 m), and winters between 1000 to 2100 m (occasionally 2700 m).

Habits and Habitat: Found in flocks of upto 20 birds, except as pairs in breeding season. Feeds among bushes but oftenly ascends trees. Prefers open forest of fir and birch with dense rhododendron and ringal bamboo growth, as well as various types of dense jungle. Feeds on insects, berries and fruits. A local migrant but very common species in Khajjiar meadow.

Breeding: Season varies between April and August.

Genus: *Heterophasia* Blyth

1842. *Heterophasia* Blyth, *Jour. Asiat. Soc. Bengal* 11: 186

47. *Heterophasia capistrata* (Vigors, 1831) (Rufous Sibia)

1831. *Cinclosoma capistratum* Vigors, *Proc. Zool. Soc. London*: 56

Size: Bulbul +; 21 cm

Diagnostic Characters: A black-capped bird, having rufous nape and underparts; slaty wings with a black shoulder patch. Tail long, graduated having slaty-grey tipped black-band.

Status and Distribution: Resident, showing vertical movements. Found in summer between 1800 to 3500 m (Occasionally from 1600 m) and in winter between 1000 to 2750 m in Himalayas.

Habits and Habitat: Moves in canopy or branches in the form of pairs or in small groups, usually in mixed evergreen forests of oak, fir, chestnut etc. Feeds on insects and berries. Often visits flowers of silk cotton trees and rhododendron for insects and nectar. This is a very common and local migrant species in Khajjiar area.

Breeding: Season varies between April and August.

Genus: *Phylloscopus* Boie

1826. *Phylloscopus* Boie, *Isis von Oken*, col. 972

48. *Phylloscopus collybita* (Vieillot, 1817) (Common Chiffchaff)

1817. *Sylvia collybita* Vieillot, *Nouv. Dict. d'Hist. Nat., nouv. ed.*, 11: 235

Size: Sparrow -; 10 cm.

Diagnostic Characters: A small, pale olive-brown bird without wing-bars; with a small whitish supercilium; olive-green edges to wing-coverts, remiges and rectrices; dull whitish underparts with buff on breast and flanks.

Status and Distribution: Winter visitor to the Northern and Central India, up to 2100 m altitude.

Habits and Habitat: Seen singly or in small parties of 8 to 10 birds, flitting restlessly from bush to bush or hopping on the ground mostly in bushes of *Acacia, Zizyphus* and some waterside vegetation. Feeds exclusively on insects like weevils, other small beetles, aphids, dipterans and geometrid larvae. Reported as winter visitor and common species from Khajjiar lake area.

49. *Phylloscopus trochiloides* (Sundevall, 1837) (Greenish Leaf-Warbler)

1838. *Acanthiza trochiloides* Sundevall, *Fysiogr. Sallskap. Tidskr. Lund.* 1: 76

Size: Sparrow -; 10 cm.

Diagnostic Characters: A small greenish-brown warbler, with a yellowish supercilium and a fine wing-bar. No crown stripe present, unlike most of other *Phylloscopus* warblers. Underparts whitish, tinged with yellow.

Status and Distribution: Summer (Breeding) visitor to Himalayas, between 2700 to 3700 m, and widespread throughout India in winter.

Habit and Habitat: Found usually singly, foraging in the foliage canopy but also in undergrowth in gardens orchards and open forests. Feeds on insects including small beetles and caterpillars. This is an uncommon and local migrant species in Khajjiar area.

Breeding: Season varies between May to August.

Genus: *Seicercus* Swainson

1837. *Seicercus* Swainson, *Classif. Bds.* 2: 84, 259

50. *Seicercus burkii* (Burton, 1836) (Gold-spectacled Flycatcher-Warbler)

1836. *Sylvia burkii* Burton, *Proc. Zool. Soc. London*: 153

Size: Sparrow -; 10 cm.

Diagnostic Characters: An olive-green warbler having two black stripes on head. A yellow eye ring conspicuous. Underparts deep yellow.

Status and Distribution: Breeds in Himalayas and Northeastern Indian hills between 2000 to 3700 m and winters down to foothills.

Habits and Habitat: Mostly seen in low undergrowth and middle storey in forests. Feeds exclusively on insects. This bird is a summer visitor and uncommon species in Khajjiar area.

Breeding: Season varies between May to July.

Genus: ***Muscicapa*** **Brisson**

1760. *Muscicapa* Brisson, *Orn.* 1: 32, 2: 357

51. *Muscicapa sibirica* Gmelin, 1789 (Sooty Flycatcher)

1789. *Muscicapa sibirica* Gmelin, *Syst. Nat.* 1 (2): 936

Size: Sparrow -; 13 cm.

Diagnostic Characters: A grey-brown flycatcher with a pale eye ring. Hawks midges by characteristic looping flights. Whitish throat patch and a white line down the centre of belly; rest of underparts grey-brown. Juvenile has spotted underparts and finely marked breast.

Status and Distribution: Breeds in Himalaya and Northeast India, winters locally throughout India.

Habits and Habitat: Mostly found perched singly on the lower branches of trees in temperate and sub-alpine forest. Feeds on insects, mostly dipterans. This is a summer visitor and common species to Khajjiar area.

Breeding: Season varies between May to July.

Genus: ***Ficedula*** **Koch**

1816. *Ficedula* Koch, *Syst. baier. Zool.* 1: 158

52. *Ficedula westermanni* (Sharpe, 1888) (Little Pied Flycatcher)

1888. *Digenea westermanni* Sharpe, *Proc. Zool. Soc. London*: 246

Size: Sparrow -; 10 cm.

Diagnostic Characters: Male has black upperparts and whitish underparts, with a conspicuous white supercilium; a large white wing patch and white sides to the basal half of tail. Female has greyish-brown upperparts; brownish grey on breast and little rufous on rump.

Status and Distribution: Resident, Himalaya and Northeast India. Recorded as a summer visitor to parts of Himachal Pradesh, between altitudes of 1200 to 3000 m.

Habits and Habitat: Usually found singly, in pairs, or in small group among the canopy of trees in wooded forest areas, orchards, etc. Consumes all type of insects. This is a common and summer visitor bird species of Khajjiar area.

Breeding: Breeds from April to June.

53. *Ficedula superciliaris* (Jerdon, 1840) (Ultramarine Flycatcher)

1840. *Muscicapa superciliaris* Jerdon, *Madras, Jour. Lit. Sci.* 11:16

Size: Sparrow -; 10 cm

Diagnostic Characters: Male has deep blue upperparts, and deep blue sides of head and neck; throat, breast and belly, white. A white supercilium and white patch to the sides of basal end of tail are conspicuous. Female is brownish, has rufous tail; greyish white breast and rest of underparts white.

Status and Distribution: The bird breeds in Himalayas and Northeast India between 1800 to 3000m, and winters south to Southern India. Feeds on insects.

Habits and Habitat: Usually moves and feeds lonely in the low trees and bushes of the forest. Ultramarine Flycatcher is a summer visitor and common species in and around Khajjiar lake glade

Breeding: Breeds between April and July.

Genus: *Eumyias* Cabanis

1851. *Eumyias* Cabanis, *Mus. Hein.* 1: 53

54. *Eumyias thalassina* (Swainson, 1838) (Verditer Flycatcher)

(Plate 19 E)

1838. *Muscicapa thalassina* Swainson, *Nat. Library, Flycatchers* 21: 252

Size: Sparrow ±; 15 cm

Diagnostic Characters: Male is entirely greenish-blue, with black lores. Female duller and greyer, with dusky lore. Juvenile is turquoise on upperparts, and the underparts with buffish spottings.

Status and Distribution: Summer visitor to Himalayas and Northeast India, between 1200 to 3000 m, and found throughout India in winters.

Habits and Habitat: Usually seen singly or in pairs, in open forest and wooded areas, perching upright and making small aerial flight for food from tips of tall tree as well as bushes. Feeds chiefly on tiny winged insects. This is a summer visitor and common species in Khajjiar area.

Breeding: Season varies from March to June.

Genus: *Niltava* Hodgson

1837. *Niltava* Hodgson, *Ind. Rev.* 1 (12): 650

55. *Niltava sundara* (Hodgson, 1837) (Rufous-bellied Niltava)

1837. *Niltava sundara* Hodgson, *Ind. Rev.* 1 (12): 650

Size: Sparrow; 15 cm

Diagnostic Characters: Male has bright ultramarine-blue crown, neck patch, shoulders, rump and tail; rest of the upperparts, black. Throat blackish; rest of underparts orange. Female is brownish with a blue patch on side of neck; fulvous brown wings; rufous brown rump and rusty brown tail. Underparts olive-brown with an oval shaped whitish throat patch.

Status and Distribution: Resident with some altitudinal movements, in Himalayas and Northeast India, found between 1600 to 2700 m.

Habits and Habitat: Found usually singly, in low undergrowth, in dense forest, close to nullahs. Feeds on insects like ants and beetles. Also feeds on berries especially in non-breeding season. Found to be a summer visitor and very common species in Khajjiar area.

Breeding: Breeds between April and July.

Family: Aegithalidae (Long-tailed Tits)

Bill very short and stout; culmen strongly curved; nostrils hidden by small feathers.

Genus: ***Aegithalos*** **Hermann**

1804. *Aegithalos* Hermann, *Obs. Zool.*: 214

First primary well developed but shorter than half the length of second. Tail longer than wing, much graduated.

56. *Aegithalos concinnus* (Gould, 1855) (Red-headed Tit)

(Plate 19 F)

1855. *Psaltoria concinna* Gould, *Birds of Asia* 2 (7): 65

Size: Sparrow -; 10 cm

Diagnostic Characters: Chestnut crown; a black strip from lores to ear coverts; grey back. Tail with white outer edges and tip. Chin and sides of black-throat white, and rest of underparts buffish.

Status and Distribution: Resident, shows some seasonal movements. Found in Himalayas and Northeast Indian hills. Found in summer between 1400 to 2700 m and winters from 600 to 3600 m.

Habits and Habitat: Found in flocks in bushes, actively investigating leaves, in broad leaved and mixed forests, and secondary growth. Feeds chiefly on in insects. Also consumes tiny seeds and fruits, particularly found of wild rasp berries. This bird is a local migrant and common species in Khajjiar area.

Breeding: Season varies from March to May.

Family: Paridae (Tits)

Bill short, conical; nostrils concealed by bristles. Rectal bristles short. Tarsus well developed, scutellated. Wing weak and rounded. Tail shorter than wing.

Genus: *Parus* Linnaeus

1758. *Parus* Linnaeus, *Syst. Nat.*, ed. 10, 1: 189

57. *Parus melanolophus* Vigors, 1831 (Spot-winged Crested Tit)

1831. *Parus melanolophus* Vigors, *Proc. Zool. Soc. London*: 23

Size: Sparrow -; 11 cm.

Diagnostic Characters: A small, greyish and black-crested bird, with white cheeks, ear coverts and a patch on hind neck, contrasting with black crown, crest and sides of neck. Two wing-bars of whitish spots, conspicuous. Throat and breast black, under tail-coverts rufous and belly grey.

Status and Distribution: Resident in Western Himalayas up to Western Nepal and shows seasonal altitudinal movements. Breeds between 2000 m to timber line (3600 m) and descends in winters up to foothills.

Habits and Habitat: Hunts restlessly in high canopy of trees, occasionally descends down to ground to pick seeds, in coniferous and mixed forests. Feeds chiefly on insects but also take seeds and berries. This is a common species with local movements in Khajjiar area.

Breeding: Season varies from May to June.

58. *Parus major* Linnaeus, 1758 (Great Tit)

1758. *Parus major* Linnaeus, *Syst. Nat.*, ed. 10, 1: 189

Size: Sparrow ±; 13 cm.

Diagnostic Characters: Head black; a white path on hind neck; greyish back; a white wing-bar on dark brown wings. Blackish tail with white outer rectrices. A black band through middle of underparts running from chin, throat and breast, connected to black of crown (enclosing white cheek patch); sides of breast and flanks, greyish-white.

Status and Distribution: Resident, shows some seasonal-local movements (especially in Himalayan regions), found almost throughout India, mostly in hilly terrain.

Habits and Habitat: Found in pairs or in small flocks, exploring the leaves and branches of trees and bushes, in forest and well wooded areas. Occasionally enters the verandas of houses. Feeds on insects, caterpillars, seeds, flower buds and berries. This is a resident and very common species in Khajjiar area.

Breeding: Season varies from April to July.

59. *Parus monticolus* Vigors, 1831 (Green-backed Tit)

1831. *Parus monticolus* Vigors, *Proc. Zool. Soc. London*: 22

Size: Sparrow ±; 13 cm

Diagnostic Characters: A bird, very similar in pattern to Great Tit *i.e.* black head, chin, throat, breast and a line through middle of the underparts; white cheeks and mantle; distinguished by green mantle and back, yellow underparts and double wing-bar.

Status and Distribution: Resident and altitudinal migrant. Found in Himalayas and Northeast Indian hills. Breeds between 1200 to 3600 m, and in winters reaches down to foothills.

Habits and Habitat: Keeps in small groups (pairs in breeding season); feeds amongst foliage of trees (occasionally ground) in forests. Feeds an insects, flowers buds fruits and berries. This is a common species of bird with winter influx in Khajjiar area.

Breeding: Season varies from March to July.

Family: Sittidae (Nuthatches)

Bill of moderate length, nearly straight, stout, compressed at the tip; wings moderate; tail short; toes long and slender; outer-toe longest, syndactyle.

Genus: ***Sitta*** **Linnaeus**

1758. *Sitta* Linnaeus, *Syst. Nat.*, ed. 10, 1: 115

Bill moderate, straight; wings rather long

60. *Sitta leucopsis* Gould, 1850 (White-cheeked Nuthatch)

1850. *Sitta leucopsis* Gould, *Bds. Asia* 2 (1), pl. 46

Size: Sparrow -; 12 cm.

Diagnostic Characters: Crown and nape black; face white and rest of the upperparts blue-slaty. Underparts, whitish with rufous lower flanks and under tail-coverts.

Status and Distribution: Resident in Western Himalayas; shows seasonal vertical movements. Breeds between 2100 to timber line and descends to 1800 m in winter.

Habits and Habitat: Mostly keeps to the tops of taller conifers, hence, difficult to see, but its presence usually detected by call. Feeds mainly on insects, also kernels of nuts and seeds. Seen as a local migrant and common species in Khajjiar areas.

Breeding: Season varies from end of April to June.

Family: Certhiidae (Tree-Creepers)

Bill as long as the head, slender and curved downwards.

Genus: ***Certhia*** **Linnaeus**

1758. *Certhia* Linnaeus, *Syst. Nat.*, ed. 10, 1: 118

Nostrils long and narrow. Tarsus scutellated. Toes and claws very long.

61. *Certhia himalayana* Vigors, 1832 (Bar-tailed Tree-Creeper)

1832 (1831). *Certhia himalayana* Vigors, *Proc. Zool. Soc. London*: 174

Size: Sparrow -; 12 cm

Diagnostic Characters: Head and back brown, streaked with pale grey. Whitish supercilium, whitish wing bar and a buff band across the wings. Tail pointed and has dark barring. Throat white and underparts dull whitish or dirty greyish-buff.

Status and Distribution: Resident with seasonal-altitudinal movements, found in Himalayas. Breeds between 2000 m to treeline and winters below 1800m, down to foothills.

Habits and Habitat: Found singly or in small groups. Works by starting at the base of tree trunk and after attaining fair height, flies to the base of another tree trunk. Prefers forests of pine, deodar, juniper, etc. in summers, and orchards, plantations and well wooded areas in winters. Feeds on insects. Found to be common species with summer influx in Khajjiar area.

Breeding: Season varies from early April to June.

Family: Zosteropidae (White-eyes)

Bill usually short, more or less wide at the base, lengthened and slightly curved in a few, entire in some, notched in others.

Genus: *Zosterops* Vigors & Horsfield

1827. *Zosterops* Vigors & Horsfield, *Trans. Linn. Soc. London* 15: 234

Wings moderate; tail short or moderate, even or slightly rounded; tarsi short, stout; feet strong; claws moderately curved, sharp.

62. *Zosterops palpebrosus* (Temminck, 1824) (Oriental White-eye)

(Plate 20 A)

1824. *Sylvia palpebrosa* Temminck, *Pl. Col d'Ois.* 49, pl. 293, fig. 3

Size: Sparrow -; 10 cm.

Diagnostic Characters: Upperparts entirely yellow-olive; throat, breast and under tail-coverts yellow; belly whitish. A distinct white eye-ring conspicuous.

Status and Distribution: Resident and seasonal migrant; found throughout India except parts of Northwest. Breeds up to 2100 m and descends down to plains in winter.

Habits and Habitat: Moves in pairs or in small groups, from tree to tree (arboreal), occasionally comes down to ground; found in open broadleaved forest and wooded areas. Feeds on insects, caterpillars, berries, buds, seeds and nectar. Oriental White-eye is an altitudinal migrant and common species in Khajjiar area.

Breeding: Season varies from April to September.

Family: Emberizidae (Buntings)

Bill with the upper mandible typically smaller and more compresses than the lower, which is broader, equal in a few; a palatal protuberance in many; commissure usually sinuate; tail moderate, even or emarginated.

Key to the Genera

A. Well-developed crest...*Melophus*

\- No Crest ....*Emberiza*

Genus: ***Melophus*** **Swainson**

1837. *Melophus* Swainson, *Classif. Bds.* 2: 290

63. *Melophus lathami* (Gray, 1831) (Crested Bunting)

(Plate 20 B)

1831. *Emberiza lathami* Gray, *Zool. Misc.* 1: 2

Size: Sparrow; 15 cm.

Diagnostic Characters: Male, entirely black with chestnut wings and tail, and with a long, pointed black crest; wing edges and outer rectrices rufous; tail nearly square, yellowish buff; breast darkly streaked; moustachial streak dark. Female olive-brown, streaked with dark brown from above; crest shorter.

Status and Distribution: Common resident, but rather local and capricious; in the hills subject to vertical movement. Found in Himalayan foothills and hills of Northeast and Central India.

Habits and Habitat: Gregarious in winter when it keeps in small loose flocks. Feeds on the ground in stony fields, on roots in hedges and thorn thickets. This is a very common species with local movements in Khajjiar area.

Breeding: Breeds between April and August.

Genus: ***Emberiza*** **Linnaeus**

1758. *Emberiza* Linnaeus, *Syst. Nat.*, ed. 10, 1: 176

64. *Emberiza cia* Linnaeus, 1766 (Rock Bunting)

1766. *Emberiza cia* Linnaeus, *Syst. Nat.*, ed. 12, 1: 310

Size: Sparrow; 15 cm

Diagnostic Characters: Male has grey head; black lateral crown-stripes; black eye-stripe and moustachial streak both meeting behind cheeks. Cheeks and throat whitish; rest of the body rufous. Outer tail-feathers white. Female, like male but duller.

Status and Distribution: Resident and altitudinal migrant, breeds between 2000 to 4600m, in Western Himalayas and winters down to adjacent plains below 2100m.

Habits and Habitat: Found singly or in small groups; feeds on the ground, occasionally perches on bushes or trees. Breeds on dry grassy and rocky slopes, and winters in cultivation. Feeds on seeds, grains and insects. This is a local migrant species which is common in Khajjiar area.

Breeding: Season varies from May to August.

Family: Fringillidae (Finches)

Bill varies in size and form, more or less conical and thick; short and bulged in some, slender and more elongate in others; wing moderate or long; first primary wanting.

Key to the Genera

A. Bill massive, upper mandible toothed near gape. Inner primaries and outer secondaries rounded at tip..................................*Mycerobas*

- Bill almost wedge-shaped, short and thick or pointed or thinner, always thick at base. Wings pointed..............................*Carduelis*

Genus: *Carduelis* Brisson

1760. *Carduelis* Brisson, *Orn.* 1: 36, 3: 53

65. *Carduelis spinoides* Vigors, 1831 (Yellow-breasted Greenfinch)

(Plate 20 C)

1831. *Carduelis spinoides* Vigors, *Proc. Zool. Soc. London*: 44

Size: Sparrow -; 14 cm.

Diagnostic Characters: It has yellow supercilium, sides of ear coverts and underparts; head, back and wings blackish brown; rump yellow; a small black malar stripe. Wings have a large yellow patch and white edged secondaries.

Status and Distribution: Resident, altitudinal migrant, Himalayas and Northeast India. Breeds between 1800 to 4000 m and winters down in foothills, below 1500 m.

Habits and Habitat: Feeds more in bushes than on the ground, in pairs or small flocks, in open forests, shrubberies and cultivation. Feeds on seeds of many kinds, including wild, sunflower, millet buckwheat, rice etc., occasionally also on berries and insects. This bird is a very common and local migrant species in Khajjiar area.

Breeding: Breeds between June to October.

Genus: *Mycerobas* Cabanis

1847. *Mycerobas* Cabanis, *in* Weigm. *Archiv. f. Naturg*. 13 (1): 350

66. *Mycerobas icterioides* (Vigors, 1831) (Black-and-Yellow Grosbeak)

1831. *Coccothraustes icterioides* Vigors, *Proc. Zool., Soc. London*: 8

Size: Myna; 22 cm.

Diagnostic Characters: Male has massive bill; black head, throat, wings, thighs and tail; collar, centre of back, rump and underparts yellow. Female has pale grey head, mantle and breast, and buff belly.

Status and Distribution: Resident and altitudinal migrant; found in Western Himalayas. Breeds between 1800 to 3500 m and descends down to 1500 m in winters.

Habits and Habitat: Usually found in pairs or small separated flocks; keeps to tops of high trees in coniferous forests, and freely descends to the ground or low bushes for feeding. Feeds on berries and small fruits, pine and crabapple seeds, fresh pine shoots. Also take insects particularly during breeding season. This is generally a local migrant bird and uncommon species in Khajjiar area.

Breeding: Season varies from April to July.

Family: Passeridae (Sparrows)

Bill stout and strong, somewhat turned, slightly compressed towards the tip; the culmen broad, convex; commissure straight; wings moderate; the first three primaries about equal, the fourth nearly as long; tail moderate nearly square, or slightly forked; tarsus moderate; feet formed both for hopping on the ground and perching; lateral toes about equal.

Genus: *Passer* Brisson

1760. *Passer* Brisson, *Orn.* 1: 36, 3: 71

Bill short and stout, culmen slightly curved.

67. *Passer domesticus* (Linnaeus, 1758) (House Sparrow)

(Plate 20 D)

1758. *Fringilla domestica* Linnaeus, *Syst. Nat.*, ed. 10, 1: 183

Size: Bulbul -; 15 cm

Diagnostic Characters: A rufous-brownish sparrow, with short-stout bill. Male has grey crown, chestnut coloured upper back and sides of crown, and rest of back rufous-chestnut with black streaks; throat and upper breast black, and sides of throat white. A prominent white shoulder patch on wing. Tail dark brown with greyish-brown rump. In winter, whitish feathers appear in throat and upper breast. Female has a pale supercilium and is greyish brown streaked with dark brown on back.

Status and Distribution: Resident, found throughout India. Recorded as local migrant in Himachal Pradesh.

Habits and Habitat: It prefers cities, towns, villages and human habitations of all kind and found in close association with man. In winters, also found in cultivation, thorn jungles, etc. far from human habitations. Feeds mostly on seeds and cereal grains, also on fruits and flower buds, tender shoots, kitchen scrap and insects. This bird is a resident and very common species in Khajjiar area.

Breeding: Breeds chiefly from March to August.

68. *Passer rutilans* Temminck, 1835 (Cinnamon Tree Sparrow)

1835. *Passer rutilans* Temminck, *Planch. Color. d'Ois.* 3, pl. 488

Size: Sparrow; 15 cm.

Diagnostic Characters: Male has bright chestnut crown and back, with black streakings on back; two white bars on wing; chin and centre of throat black; rest of underparts yellowish. Female has brown upperparts streaked with dark; prominent supercilium; yellowish wash to underparts.

Status and Distribution: Resident and seasonal-altitudinal migrant in Himalayas and Northeast India hills. Breeds between 1200 m to 2700 m and winters between 500 m to 1500 m.

Habits and Habitat: Feeds in flocks on grain and grass seeds, also usually seen perched on exposed places like trees or wires especially in hill stations. Preferred habitat is open forest, forest edges and cultivation. Feeds chiefly on grains seeds, also on berries and insects. A very common species with summer influx from Khajjiar area.

Breeding: Two broods reared from April to August.

Family: Sturnidae (Mynas)

Bill straight, or very slightly curved, longish, compressed, subulate, often angulated at the base, slightly notched at the tip or entire; wings long, rather pointed; tail moderate or stout; tarsi short, moderate; lateral toes about equal.

Genus: ***Acridotheres*** **Vieillot**

1816. *Acridotheres* Vieillot, *Anal. nouv. Orn.*: 42

A white wing-patch on the upperparts, near base of remiges. Underparts and throat mostly dark grey.

69. *Acridotheres tristis* (Linnaeus, 1766) (Common Myna)

(Plate 20 F)

1766. *Paradisea tristis* Linnaeus, *Syst. Nat.*, ed. 12, 1: 167

Size: Dove -; 23 cm

Diagnostic Characters: A familiar, brownish bird with black head, bright yellow legs and bill, and a bright yellow patch below legs. A large white patch on wing, conspicuous in flight. Young is paler than adult.

Status and Distribution: Resident, throughout India.

Habits and Habitat: A commensal of man, along with crows and sparrows. It is an important agent in cross pollination of some species of flowers and in the dispersal of seeds of many plant species. Also causes damage to crops and orchards etc. Feeds mainly on fruits, grains, insects and grubs, also on everything else that can be eaten like kitchen scraps, titbits from refuse dumps, small animals such as baby mice, frogs, laggards, crabs and flowers nectar. Common Myna is a resident and very common species in Khajjiar area.

Breeding: Season varies between March and September.

70. *Acridotheres fuscus* (Wagler, 1827) (Jungle Myna)

1827. *Pastor fuscus* Wagler, *Syst. Av. Pastor* sp. 6

Size: Myna; 23 cm.

Diagnostic Characters: Overall bluish grey and black, with a tuft of feathers on forehead and reddish to orange-brown iris (without naked orbital skin). Wing-patch, tail tip and vent white. Juvenile browner than adult.

Status and Distribution: Resident, Himalayas south to Northern India, west Bengal and Northern Orissa, upto 2400m in Himalayas.

Habits and Habitat: Found in family parties or flocks of 10 to 30 birds, around cultivation near well-watered areas, and edges of habitation. Feeds on fruits and berries, grains, flower nectar and insects. This bird is a resident and common species in Khajjiar area.

Breeding: Double brooded; season varies from February to July.

Family: Dicruridae (Drongos)

Bill rather large, wide at the base, thick, more, or less curved and keeled at the culmen, and notched at the tip; numerous moderately strong rictal bristles; nostrils basal, rounded, concealed by short plumes; wings lengthened; fourth and fifth quills usually the largest.

Genus: *Dicrurus* Vieillot

1816. *Dicrurus* Vieillot, *Anal. Nouv. Orn.*: 41

Legs short; feet small; tail usually long, forked; the outer feathers occasionally much lengthened; of ten feathers.

71. *Dicrurus macrocercus* Vieillot, 1817 (Black Drongo)

1817. *Dicrurus macrocercus* Vieillot, *Nouv. Dict. d'Hist. Nat.*, nouv. ed., 9: 588

Size: Bulbul +; 28-31 cm.

Diagnostic Characters: A glossy blue-black, fork-tailed bird. Adult has white rectal spots on blue-black underparts. Tail fork may be lost during moulting. First winter birds have bold whitish firings on underparts. Juvenile has dark brown appearance.

Status and Distribution: Resident throughout India, reaching up to 2100 m altitude in Himalayas, shows seasonal local and altitudinal movements.

Habits and Habitat: A great friend of agriculture due to its complete carnivorous habits, as it feeds on a large variety of insect pests. Usually seen singly, perched on leafless tree-tops, wires, fence posts, etc., around open forest, cultivated country and outskirts of habitation. Feeds predominantly on insects, at occasions on small birds, and small bats. This bird is considered a resident and very common species in Khajjiar area.

Breeding: Season varies from March to June/July, varying locally.

Family: Corvidae (Crows, Jays, Treepies, Magpies)

Bill strong, more or less compressed, usually entire, rarely notched at the tip; nostrils thickly clad with stiff incumbent bristles; tarsus short; feet strong, and claws well curved; of large size mostly.

Key to the Genera

A. Plumage predominantly black, or black with ashy nape.........*Corvus*

B. Plumage with brilliant or contrasting colours

   1. Small white patches on primaries

      a. Rectrices not elongated.................................*Garrulus*

b. Two central rectrices elongated

i. Plumage with shades of blue.....................*Urocissa*

ii. Plumage without blue..........................*Dendrocitta*

Genus: *Garrulus* Brisson

1760. *Garrulus* Brisson, *Orn.* 1: 30, 2: 47

72. *Garrulus glandarius* (Linnaeus, 1758) (Eurasian Jay)

1758. *Corvus glandarius* Linnaeus, *Syst. Nat.*, ed. 10, 1: 105

Size: Dove; 32-36 cm.

Diagnostic Characters: A pinkish to Reddish-brown bird, with a black moustachial stripe; blue and blackish wings; white rump; black tail.

Status and Distribution: Resident, Himalaya and Northeast India. Found between 1500 to 3000 m, occasionally down to 1000 m.

Habits and Habitat: Found in noisy pairs or in small groups in association with Jays and Treepies, mainly in broadleaved temperate forest. Feeds on fruits and nuts, insects, lizards, small mammals, eggs of other birds. A very common and local migrant bird species in Khajjiar area.

Breeding: Breeds mainly in April and May.

73. *Garrulus lanceolatus* Vigors, 1831 (Black-headed Jay)

1831. *Garrulus lanceolatus* Vigors, *Proc. Zool. Soc. London*: 7

Size: Dove; 33 cm.

Diagnostic Characters: Overall vinous-grey with black face and crest; streaked throat and blue barring on wings and tail.

Status and Distribution: Resident, Northwest Himalayas, found between 900 to 2500 m. **Habits and Habitat:** Found in noisy pairs or in small parties, in mixed temperate forest. Feeds on fruits and nuts, insects, lizards, small mammals, eggs of other birds. Reported a local migrant and uncommon species in Khajjiar areas.

Breeding: Breeds mainly during May.

Genus: *Urocissa* Cabanis

1851. *Urocissa* Cabanis, *Mus. Hein.*, pt. 1: 87.

74. *Urocissa flavirostris* (Blyth, 1846) (Yellow-billed Blue Magpie)

1846. *Ps. (ilorhinus) flavirostris* Blyth, *Jour. Asiat. Soc. Bengal* 15: 28

Size: Pigeon ±; 61-66 cm

Diagnostic Characters: A purplish-blue bird with black head, neck and breast; long graduated white-tipped tail; white underparts; yellow bill and a white crescent patch on nape.

Status and Distribution: Resident, Himalayas and Northeast India. Found between 1600 to 2700 m during summer and descends down to 800m during winter. Occupy higher altitudinal zone than Red-billed Blue Magpie.

Habits and Habitat: Found in noisy flocks in association with jays and laughing thrushes, in temperate mixed forests. Feeds mainly on insects, frogs, lizards, small snakes, eggs, birds and small mammals, also on fruits, barriers and kitchen scrap. This is a local migrant and uncommon species in Khajjiar area.

Breeding: Breeding season varies from May to June.

75. *Urocissa erythrorhyncha* (Boddaert, 1783) (Red-billed Blue Magpie)

(Plate 20 G)

1783. *Corvus erythrorhyncha* Boddaert, *Tabl. Pl. Enlum.*: 38

Size: Pigeon ±; 70 cm

Diagnostic Characters: A brightly-blue coloured, long-graduated tailed bird, with red bill. Head, neck and underparts, behind breast greyish-white. A white patch on the nape.

Status and Distribution: Found in Himalayas and Northeast India, from Kangra eastwards, between foothills to 2100 m and shows some summer-winter altitudinal movements.

Habits and Habitat: The bird is found in noisy parties of 4 to 10 individuals, in well wooded areas. Mainly leads arboreal life but descends to the ground to feed upon insects, fallen fruits or grains in harvested fields. Feeds mainly on insects, frogs, lizards, small snakes, eggs, birds and small mammals, also on fruits, barriers and kitchen scrap. This is a common bird species with summer influx in Khajjiar area.

Breeding: Breeds mainly from April to June.

Genus: *Dendrocitta* Gould

1833. *Dendrocitta* Gould, *Proc. Zool. Soc. London*: 57

76. *Dendrocitta formosae* Swinhoe, 1863 (Grey Treepie)

1863. *Dendrocitta formosae* Swinhoe, *Ibis*: 387

Size: Myna ±; 43 cm

Diagnostic Characters: A long tailed, grey and brown treepie, with dark-grey face; grey underparts and rump; black wings with a pure white patch. Under tail-coverts chestnut in colour.

Status and Distribution: Resident, found in Himalayas, Northeast India and Eastern Ghats, up to 2100 m altitude.

Habits and Habitat: Usually seen in small parties of 4 or 5 birds or more. It prefers well wooded areas and edges of cultivation. Mostly prefers trees for feeding but can be seen in harvested fields also. Feeds on fruits, seeds, flower nectar, insects, centipedes, lizards, eggs young of small birds and ants small animals. Grey Treepie is a resident and common species in Khajjiar area.

Breeding: Breeds from April to July between 700 and 1600 m altitude.

A Cattle Egret

B Egyptian Vulture

C Himalayan Griffon

D Himalayan Monal

E Kaleej Pheasant

F Blue Rock Pigeon

A Wedge-tailed Green Pigeon

B Plum-headed Parakeet

C White-breasted Kingfisher

D Common Hoopoe

E Great Barbet

F Long-tailed Minivet

Plate 17: Some birds recorded in Khajjiar lake area

A Himalayan Bulbul

B Black Bulbul

C Bay-backed Shrike

D Chestnut-bellied Rock Thrush

E Blue Whistling Thrush

F White-capped Redstart

Plate 18: Some birds recorded in Khajjiar lake area

A Spotted Forktail

B Common Stonechat

C Grey Bushchat

D Streaked Laughing Thrush

E Verditer Flycatcher

F Red-headed Tit

Plate 19: Some birds recorded in Khajjiar lake area

A. Oriental White-eye

B. Crested Bunting

C Yello-breasted Greenfinch

D House Sparrow

E Russet Spaarrow **F Common Myna**

G Red-billed Blue Magpie **H Large-billed Crow**

Plate 20: Some birds recorded in Khajjiar lake area

Genus: *Corvus* Linnaeus

1758. *Corvus* Linnaeus, *Syst. Nat.*, ed. 10, 1: 105

77. *Corvus macrorhynchos* Wagler, 1827 (Jungle Crow)

(Plate 20 H)

1827. *Corvus macrorhynchos* Wagler, *Syst. Av.*, *Corvus*, sp. 3

Size: House Crow +; 50 cm

Diagnostic Characters: A large, uniformly black crow with heavy black-bill, and lacks paler collar of house crow.

Status and Distribution: Resident, found almost throughout India, upto 4500 m in Himalayas, and also recorded up to 6400 m (Everest).

Habits and Habitat: It is absent from desert and semi desert areas and loves well-wooded country, outskirts of forest villages and rural areas, as well as towns and cities. In summers, follows herds of sheep and goats in small parties or flocks of up to 50 birds. Feeds on a variety of food, from animal to plant origin. This bird is very common and resident species in Khajjiar area.

Breeding: Breeds from March to May.

V. MAMMALS

16 species of mammals belonging to 14 genera, 12 families and 6 orders are present in the Khajjiar wildlife sanctuary. Order wise analyses of the data reveals that order Carnivora supports maximum of 6 species followed by Artiodactyla (3 species), and Chiroptera, Primates and Rodentia (2 species each). Moreover a single species belonging to order Insectivora is also present (Table 5.2). Family wise analysis of data reveals that families Vespertilionidae, Cercopithecidae, Mustelidae and Bovidae are represented by two species each whereas, Ursidae, Hynaeidae, Felidae, Cervidae, Sciuridae, Muridae, Soricidae and Canidae are represented by one species each. Of the sixteen species of mammals, 11 species belongs to the category of large mammals and 5 species to the category of small mammals.

Nine species are listed as threatened in Convention in Trade of Endangered Species (CITES) under different schedules. Five species namely *Semnopithecus ajax*, *Ursus thibetanus*, *Panthera pardus*, *Naemorhedus sumatraensis* and *Naemorhedus goral* are placed in schedule I, *Macaca mulatta* in Schedule II and *Vulpes vulpes*, *Martes flavigula* and *Mustela sibrica* under schedule III.

Out of a total of sixteen species thirteen are placed under Indian Wildlife Protection Act 1972. Two species *Panthera pardus* and *Naemorhedus sumatraensis* are kept under schedule I. Same species is considered as vulnerable species according to National Red Data.

Mammals are reported from all three parts of Khajjiar i.e. dense forest, human settlement and lake meadow of the khajjiar area. Species like Leopard, Yellow Throat Marten, Himalayan Weasel, Black Bear, Striped Hyena, Barking Deer, Flying Squirrel, Himalayan Fox, Serow and Goral can be seen in the forest area. Himalayan Fox can be observed from the grassy slopes outside the lake on village footpaths. Himalayan

Table 5. 1: Systematic list of birds and their habits in Khajjiar lake area

S.No.	*Taxon*	*R. A.*	*R. S.*	*F.H.*
	Order: Ciconiiformes			
	Family: Ardidae			
1.	Cattle Egret- *Bubulcus ibis* (Linnaeus, 1758)	R/LM	VC	qA
	Order: Falconiformes			
	Family: Accipitridae			
2	Black Kite -*Milvus migrans* (Boddaert, 1783)	R	VC	OM
3	Bearded Vulturei –*Gypaetus barbatus* (Linnaeus, 1758)	R	C	SC
4	Egyptian Vulture - *Neophron percnopterus* (Linnaeus, 1758)	R/LM	C	SC
5	Himalayan Griffon -*Gyps himalayensis* (Hume, 1869)	R	VC	SC
6	Cinereous Vulture -*Aegypius monachus* (Linnaeus, 1766)	WV	Ra	SC
7.	Black Eagle - *Ictinaetus malayensis*(Temminck, 1822)	R/LM	UC	CR
	Order: Galliformes			
	Family: Phasianidae			
8.	Koklass Phesant- *Pucrasia macrolopha*(Lesson, 1829)	R/LM	C	OM
9.	Impeyan Monal -*Lophophorus impejanus* (Latham, 1790)	R	UC	gM
10.	Kaleej Pheasant -*Lophura leucomelanos* (Latham, 1790)	R	C	GR
	Family: Recurvirostridae			
11.	Black-winged Stilt- *Himantopus himantopus* (Linnaeus, 1758)	UC	OM	SV
	Order: Columbiformes			
	Family: Columbidae			
12	Blue Rock Pigeon -*Columba livia* Gmelin, 1789	R	VC	GR
13.	Oriental Turtle-Dove- *Streptopelia orientalis* (Latham, 1790)	R/LM	VC	GR
14.	Wedge-tailed Green-Pigeon-*Treron sphenura* (Vigors, 1832)	R/SV	C	FR
	Order: Psittaciformes			
	Family: Psittacidae			
15.	Slaty-headed Parakeet -*Psittacula himalayana* (Lesson, 1832)	R/LM	VC	FR
16.	Plum-headed Parakeet- *Psittacula cyanocephala*	R/SV	C	FR

Contd.

Table 5.1: Contd....

S.No.	Taxon	R.A.	R.S.	F.H.
	Order: Cuculiformes			
	Family: Cuculidae			
17.	Common Cuckoo -*Cuculus canorus* Linnaeus, 1758	R/LM	VC	I
	Family: Strigidae			
18.	Asian Barred Owlet- *Glaucidium cuculoides* (Vigors, 1831)	R/LM	C	ICR
	Order: Caprimulgiformes			
	Family: Caprimulgidae			
19.	Large-tailed Nightjar-*Caprimulgus macrurus* Horsfield, 1821	SV	C	I
	Order: Apodiformes			
	Family: Apodidae			
20.	House Swift - *Apus affinis* (J.E. Gray, 1830)	R	C	I
	Order: Coraciiformes			
	Family: Alcedinidae			
21.	White-breasted Kingfisher - *Halcyon smyrnensis*	R	VC	qA, I
22.	Greater Pied Kingfisher- *Megaceryle lugubris* (Temminck, 1834)	R	C	AqA
	Family: Upupidae			
23.	Common Hoopoe -*Upupa epops* Linnaeus, 1758	WV	C	I
	Order: Piciformes			
	Family: Capitonidae			
24.	Great Barbet - *Megalaima virens* (Boddaert, 1783)	R/LM	UC	FR,I
	Family: Picidae			
25.	Himalayan Pied Woodpecker- *Dendrocopos himalayensis* (Jardine and Selby, 1831)	R/LM	VC	I
26.	Scaly-bellied Green Woodpecker-*Picus squamatus* Vigors, 1831	R	UC	I
27.	Black-naped Green Woodpecker - *Picus canus* Gmelin, 1788	R/LM	UC	I
	Order: Passeriformes			
	Family: Hirundinidae			
28.	Red-rumped Swallow - *Hirundo daurica* Linnaeus, 1771	R/SV	C	I

Contd.

Table 5.1: Contd....

S.No.	Taxon	R. A.	R. S.	F.H.
	Family: Motacillidae			
29.	Large Pied Wagtail- *Motacilla maderaspatensis* Gmelin, 1789	R/LM	UC	I
	Family: Campephagidae			
30.	Long-tailed Minivet- *Pericrocotus ethologus* Bangs and Phillips, 1914	SV	C	I
	Family: Pycnonotidae			
31.	Himalayan Bulbul -*Pycnonotus leucogenys* (Gray, 1835)	R	VC	FR
32.	Black Bulbul - *Hypsipetes leucocephalus* (P.L.S. Muller, 1776)	R	C	FR
	Family: Laniidae			
33.	Bay-backed Shrike - *Lanius vittatus* Valenciennes, 1826	WV	UC	CR
	Family: Muscicapidae			
34.	Blue-headed Rock-Thrush- *Monticola cinclorhynchus* (Vigors, 1832)	SV	UC	I
35.	Chestnut-bellied Rock-Thrush - *Monticola rufiventris* (Jardine & Selby, 1833)	R/LM	UC	I
36.	Blue Whistling-Thrush - *Myiophonus caeruleus* (Scopoli, 1786)	R/LM	C	CR
37.	White-collared Blackbird- *Turdus albocinctus* Royle, 1840	R/LM	UC	I,FR
38.	Grey-winged Blackbird -*Turdus boulboul* (Latham, 1790)	R/WV	UC	I
39.	Mistle Thrush- *Turdus viscivorus* Linnaeus, 1758	R/LM	UC	I
40.	White-capped Redstart- *Chaimarrornis leucocephalus* (Vigors, 1831)	R/LM	C	I
41.	Plumbeous Redstart- *Rhyacornis fuliginosus* (Vigors, 1831)	R/LM	C	I
42.	Spotted Forktail- *Enicurus maculatus* Vigors, 1831	R/LM	UC	AqA
43.	Common Stonechat- *Saxicola torquata* (Linnaeus, 1766)	R/LM	VC	I
44.	Grey Bushchat- *Saxicola ferrea* Gray, 1846	R/LM	VC	I
45.	Streaked Laughingthrush- *Garrulax lineatus* (Vigors, 1831)	R/LM	VC	I
46.	Variegated Laughingthrush- *Garrulax variegatus* (Vigors, 1831)	R/LM	VC	I, FR
47.	Rufous Sibia- *Heterophasia capistrata* (Vigors, 1831)	R/LM	VC	I

Contd.

Table 5.1: Contd....

S.No.	Taxon	R. A.	R. S.	F.H.
48.	Common Chiffchaff- *Phylloscopus collybita* (Vieillot, 1817)	WV	UC	I
49.	Greenish Leaf-Warbler- *Phylloscopus trochiloides* (Sundevall, 1837)	R/LM	UC	I
50.	Gold-spectacled Flycatcher-Warbler *Seicercus burkii* (Burton, 1836)	SV	UC	I
51	Sooty Flycatcher- *Muscicapa sibirica* Gmelin, 1789	SV	C	I
52.	Little Pied Flycatcher -*Ficedula westermanni* (Sharpe, 1888)	SV	C	I
53.	Ultramarine Flycatcher-*Ficedula superciliaris* (Jerdon, 1840)	SV	C	I
54.	Verditer Flycatcher -*Eumyias thalassina* (Swainson, 1838)	SV	UC	I
55.	Rufous-bellied Niltava -*Niltava sundara* (Hodgson, 1837)	SV	VC	I
	Family: Aegithalidae			
56.	Red-headed Tit- *Aegithalos concinnus* (Gould, 1855)	R/LM	C	I
	Family: Paridae			
57.	Spot-winged Crested Tit- *Parus melanolophus* Vigors, 1831	R/LM	C	I
58.	Great Tit- *Parus major* Linnaeus, 1758	R	VC	I
59.	Green-backed Tit- *Parus monticolus* Vigors, 1831	R/WV	C	I
	Family: Sittidae			
60.	White-cheeked Nuthatch- *Sitta leucopsis* Gould, 1850	R/LM	C	I
	Family: Certhiidae			
61.	Bar-tailed Tree-Creeper- *Certhia himalayana* Vigors, 1832	R/SV	C	I
	Family: Zosteropidae			
62.	Oriental White-eye *Zosterops palpebrosus* (Temminck, 1824)	R	C	OM
	Family: Emberizidae			
63.	Crested Bunting- *Melophus lathami* (Gray, 1831)	R/LM	VC	GR
64.	Rock Bunting- *Emberiza cia* Linnaeus, 1766	R/LM	C	GR

Contd.

Table 5.1: Contd....

S.No.	*Taxon*	*R. A.*	*R. S.*	*F.H.*
	Family: Fringillidae			
65.	Yellow-breasted Greenfinch- *Carduelis spinoides* Vigors, 1831	R/LM	VC	GR
66.	Black-and-Yellow Grosbeak- *Mycerobas icterioides* (Vigors, 1831)	R/LM	UC	FR
	Family: Passeridae			
67.	House Sparrow- *Passer domesticus* (Linnaeus, 1758)	R	VC	GR
68.	Cinnamon Tree Sparrow- *Passer rutilans* Temminck, 1835	R/SV	VC	GR
	Family: Sturnidae			
69.	Common Myna- *Acridotheres tristis* (Linnaeus, 1766)	R	VC	OM
70	Jungle Myna- *Acridotheres fuscus* (Wagler, 1827)	R	C	FR
	Family: Dicruridae			
71.	Black Drongo- *Dicrurus macrocercus* Vieillot, 1817	R	VC	OM
	Family: Corvidae			
72.	Eurasian Jay- *Garrulus glandarius* (Linnaeus, 1758)	R/LM	VC	OM
73.	Black-headed Jay- *Garrulus lanceolatus* Vigors, 1831	R/LM	UC	OM
74.	Yellow-billed Blue Magpie- *Urocissa flavirostris* (Blyth, 1846)	R/LM	UC	OM
75.	Red-billed Blue Magpie- *Urocissa erythrorhyncha*	R/SV	C	OM
76.	Grey Treepie- *Dendrocitta formosae* Swinhoe, 1863	R	C	OM
77.	Jungle Crow- *Corvus macrorhynchos* Wagler, 1827 (Boddaert, 1783)	R	VC	OM

Weasel which generally lives in forests, in open grass and scrub, take shelter amongst rocks, under roots of trees, in hollow stumps or logs and quite often in the burrow of some other animals are reported from the rocky slopes which are present on the way to Khajjiar Lake. No direct sighting of Black Bear can be made however, interview with local residents and some indirect evidences like destruction of maize crop by bear points towards the presence of good population of this species in the Khajjiar area. Likewise, Leopard is not easily sighted directly in the present area but many indirect evidences support the presence of this animal in the area. 'Kills of animals' are generally reported on different occasions and near to these kills 'Pug Marks' can be observed. Leopard is not restricted only to dense forests or heavy covers and thrives well in open country among rocks and scrubs. Primarily nocturnal, but can hunt in day time being more tolerated to sun light. Many locals complained that their pet animals have been attacked by Leopard in the night.

Order: Insectivora

Family: Soricidae (Shrews)

Small animals due to their general appearance and nocturnal habits, commonly resemble rats and mice. Body covered with small hairs. Small but distinct eyes, ears small and external. Snout lengthened, pointed and very mobile. Limbs as a rule short and five toed. Toes well cloven and moderate lengthened, more or less naked or thinly clawed with hairs.

Genus: ***Suncus*** **Ehrenberg**

1832. *Suncus* Ehrenberg, *In hemprich and Ehrenberg*, Symb. Phys. Mamm., 2: k

1. *Suncus murinus* Linnaeus, 1766 (House Shrew)

1766. *Suncus murinus* Linnaeus, *Syst. Nat. 12th ed.*, 1:74

Diagnostic Characters: A typical and large shrew with small black eyes and small external ears. Snout long pointed with profuse soft vibrissae. Body dark brown, dorsum blackish, vent grey, without clear demarcation between the two. Ear sparsely haired with a distinctive fold on conch. Tail thickened at the base and tapers to a fine point at the end with long white hairs. The upper tooth row of an incisor, four platal premolars and four molars. Legs short with five clawed toes.

Distribution: Found in Afghanistan, Pakistan, Nepal, China, Bhutan, Myanmar, Sri Lanka, continental and peninsular Indo-Malayan Region. Introduced into Guam, Maldives, Mauritius, Madagascar etc.

Habits and Habitat: Active both during day as well as night. Breed throughout the year. Normally feeds on insects and also reported to attack and eat Bull Frog (*Rana tigrina*) and even snake (*Natrix atollata*). Generally reported from Khajjiar meadow in good numbers.

Order: Chiroptera

Family: Vespertilionidae

The family comprises of ordinary bats. Animals of this family have no facial leaf and ears separated. Upper incisors four to two in numbers, the middle ones apart. Lower incisors sharp and somewhat notched, number four to six. These animals have simple lips, long tail, and wide wings. Index fingers have single phalanx.

Genus: ***Myotis*** **Kaup**

1829. *Myotis* Kaup, *Skizz. Entwickel.-Gesch. Nat. Syst. Europ. Thierwelt*, 1: 106

2. *Myotis muricola* Gray, 1846 (Nepalese Whiskered Bat)

1846. *Myotis muricola* Gray, *Cat. Hodgson Coll. Brit. Mus.*: 4

Diagnostic Characters: Upperparts brown to gray with dark bases and the underparts with dark bases, light brown tips. Ears moderately long, well-developed slender tragus, bent forwards and bluntly pointed. Feet small with wing membrane attached at the base of toes. Tail long, completely enclosed in the interfemoral membrane. Three pairs of premolars with upper canine much longer than third premolar. Wing membrane attached to the side of the foot at the base of the toes.

Distribution: Found in Indo-Malayan Region, including most of southeast Asia, east India, Indonesia and the Philippines; also in Taiwan and central southern China. In India, reported from Shimla, Chamba and Kangra districts.

Habits and Habitat: Very active during summer months and probably hibernates for a long period. Breeding period has been reported to range between May and June in Shimla. Included on the basis of published records from Dalhousie (Dodsworth, 1913).

3. *Myotis blythii* Tomes, 1857 (Lesser Mouse-eared Bat)

1857 *Myotis blythii* Tomes, *Proc. Zool. Soc. Lond.*, 53

Diagnostic Characters: A large bat with slightly raised crown. Ears large, extending to just beyond the end of the muzzle when laid forward. Dorsal colour of the adult varies from greyish brown-tinged yellow to grey-tinged dark brown. Ventral colour varies from very slightly dirty white to greyish white. White tuft of hairs between the ears present. Only extreme tip of the tail project from the interfemoral membrane.

Distribution: Found in countries like Afghanistan, Australia, China, France, Georgia, Germany, Greece, Hungry, Iran, Iraq, Italy, Israel, Kazakhstan, Mongolia, Pakistan, Romania and Russia. In Himachal Pradesh reported from Chamba, Dalhousie, Shimla and Karol hill near Solan.

Habits and Habitat: A nocturnal creature, spends most of the day hanging from roof of the narrow caves, frequently mixing with other bat species. Emerges out first of all the other species but only after darkness sets in fully. This species can be observed in a short and narrow cave at a hilltop (2200 m alt.) in the Khajjiar area.

Order: Primates

Family: Cercopithecidae

Dentition of animals of this family very closely resembles to the dentition of man differing in the incisors and especially in canines being larger. They have dense fur on the body. Face and hand devoid of hairs. Have long forearms, long flexible toes and thumb on hind feet. Forefeet often larger than hind feet. They have long tails which perhaps help them in balancing themselves in their surprisingly long leaps. Pelvis narrow, does not assist in equilibrium in erect position. They cannot stand on the soles of their feet but rest on the outer edge of foot.

Genus: *Macaca* Lacepede

1799. *Macaca* Lacepede, *Tabl. Div. Subd. Orders Genres Mammiferes*: 4

4. *Macaca mulatta* (Zimmermann, 1780) (Rhesus Monkey)

(Plate 21 A)

1780. *Ceropithecus mulatta* Zimmermann, *Georg. Gesch. Mensch.*

Vierf. Thiere. 2: 195

Diagnostic Characters: Overall appearance orange brown with brown shades varying according to the season and habitat. No fur on reddish pink or red faces. Rump reddish and tail medium sized. A slight sexual dimorphism does occur. Height and weight of males little more than females. *Locomotes on four feet and depending upon their habitat they can be arborial or terrestrial.* This species has cheek pouches to carry food in while it forages.

Distribution: Found almost everywhere in Indian Peninsula. Rhesus macaques are found ubiquitously throughout mainland Asia; from Afghanistan to India and Thailand to southern China. The only primates with a broader geographic distribution than Rhesus Macaques are humans.

Habits and Habitat: Found in a wide range of habitat, can tolerate a diverse temperature and consume a variety of food. Found in almost all kinds of forests in wild environment but also found in areas close to humans in urban settings or near cultivation, on roadsides, canal banks, in railway stations, villages, towns, and temples. This is a common species of mammal, observed in different sized groups in Khajjiar.

Conservation Status (CITES): Schedule II

Genus: ***Semnopithecus*** **Desmarest**

1822. *Semnopithecus* Desmarest, *Mammalogie, in Encycl. Meth, 2 (Supp.)*: 532

5. *Semnopithecus ajax*, Pocock, 1928

(Plate 21 C)

1928. *Semnopithecus ajax* Pocock, *J. Bombay Nat. Hist. Soc.*, 32:480.

Diagnostic Characters: larger size and outer sides of both the fore and hind limbs covered with silvery-dark colored hair Tail always longer than the body, black skin around eyes and cheeks, muzzle framed collar of radiating creamy white hairs, arms dark. Rest of the body covered with greyish fur. Tail fur also greyish ending into a tuft of long white hairs. Males slightly larger than the females.

Distribution: Found throughout India except northeast and western part of Gujarat. Elsewhere found in Pakistan, China, Nepal Sri Lanka.

Habits and Habitat: Terrestrial and live in forests, open wooden habitats and also in urban areas around human settlements. Diurnal and walk on four legs, sleep on trees. Live in groups of 18 to 25 and have a social relationship among themselves. One group is dominated by one male. Generally feed on leaves, fruits, buds and flowers but sometimes also feed on insects, tree bark and gum.

One of the commonest large mammals in Khajjiar area can be seen equally around human settlements and forest areas. Reported to damage the agricultural crops in the present study area.

Conservation Status (CITES): Schedule I

Order: Carnivora

Family: Canidae

All the animals built on same general plan, a well shaped head, long pointed muzzle, large erect ears, deep chested muscular body, bushy tail and slender limbs. Their perfectly digigrade feet have blunt, nearly straight and non retractile claws. Forefeet with five toes, the tumb raised and hind feet usually with four toes. Jaws and teeth specially designed for capturing the prey, jaw bones long and variously armed with teeth. Sufficient space left between the canines and neighbouring incisors and check teeth to allow their interlocking.

Genus: *Vulpes* Frisch

1775. *Vulpes* Frisch, *Das Nature- System der Vierfussigen Theire*: 15

6. *Vulpes vulpes* Linnaeus, 1758 (Himalayan Fox)

1758. *Vulpes vulpes* Linnaeus, *Syst. Nat.* 10th ed., 1:39

Diagnostic Characters: A richly coloured fox with large silky fur. Colouration of the fur varies from pale yellow red to deep reddish brown on upperparts and pure white, grey or silvery on underparts, the lower parts of the body and legs generally black. The black backs to the upper half's of ears and white tips of tail distinguish it from other Indian foxes. Body of the animal well proportioned with slender limbs. The head pointed and narrow with pointed and blackish ears. Length of the body 55 to 75 cm, tail length 40 to 50 cm inclusive of hairs. Average weight of the body 5 kilograms.

Distribution: Within Indian limits found in Tibet, Ladakh, Kashmir and the Himalayas as far east as Sikkim. Elsewhere: Tibetan plateau, Pakistan and Afghanistan.

Habits and Habitat: Prefers a dry rather than moist climate. Generally found in semi arid open grasslands or dry boulder-set slopes. Shelter in burrows dug in the ground, under or among rocks, and around vegetation among reeds and bushes. Although nocturnal, may be seen hunting by day in the winter. A monogamous and territorial species. The diet mainly omnivorous, feeds chiefly on rodents, lagomorphs, insects and also on fruits. Himalayan Fox pairs for life and occupy the same den years after years. This mammal species is generally observed in Khajjiar area in groups of two to four individuals.

Conservation Status (CITES): Schedule III

Order: Carnivora

Family: Mustelidae

Representative animals of this family include weasels and martens. Animals of small size, elongated vermiform make and very short limbs. Head rounded in front but distance from the orbit to occipital foramen very large, so skull has an elongated form posteriorly. Upper jaw has four to five molar teeth on each side and lower jaw has five or rarely six. Feet pentadactyle having sharp claws. Snout short and slender. Very active and highly carnivorous animals.

Genus: ***Martes*** **Pinel**

1792. *Martes* Pinel, *Actes. Soc. Hist. Paris*, 1: 55

7. *Martes flavigula* Boddaert, 1785 (Yellow throated Marten)

1785. *Martes flavigula* Boddaert, *Elench. Anim.*, 1: 88

Diagnostic Characters: A graceful slender animal having comparatively large body size and limbs which distinguish it from stone martens. Tail relatively long measuring about three fourths of length of the head and body. Colour varies among individuals and with season. It has an orange-yellow to dark brown coat, with a creamy yellow throat; chin and lower lips white. Head and body length about 40-60 cm with a tail length of 38-43 cm. Ears low-set and rounded. Feet have naked pads with sharp claws. Weight of adult marten about 3.4 kg.

Distribution: Found throughout Himalayas. Outside India found in central and eastern Asia, upper Myanmar and Java.

Habits and Habitat: Found in tropical, subtropical, temperate and sub alpine forest belts of Himalayas usually between elevations of 1200 to 2700 metres. It makes it den in any convenient shelter, among rocks, under roots of trees, in hollows of stumps and many times in burrows of other animals. They may hunt in day but more usually come after nightfall. Hunt for mice, birds and bird's egg, reptiles and insects. When near human settlements attack domestic chickens and pigeon. Normally lives and hunts alone except during mating time or when with young ones.

Conservation Status (CITES): Schedule III

Genus: ***Mustela*** **Linnaeus**

1758. *Mustela* Linnaeus, *Syst. Nat.*, 10th ed., 1: 45

8. *Mustela sibrica* (Pallas) (Himalayan Weasel)

1773. *Mustela sibrica* Pallas, *Reise. Prov.Russ.Reichs*. 2: 701

Diagnostic Characters: Weasels have long, stretched-out bodies with relatively short legs; elongated, narrow and relatively small head; ears broad at the base, but short; tail of about half of the body length. Colour varies from foxy red to dark chocolate; no sharp contrast between upper and underparts. The colour of muzzle black in some forms and more or less white in others forms. Not much distinction between the two sexes.

Distribution: Found in Himalayas from Kashmir to northwest states. Outside India found in Russia, central and eastern Asia, upper Myanmar and Java.

Habits and Habitat: In Himalayas, lives in temperate and alpine forests and in open grass and scrub, above treeline at altitudes ranging from 1500 to 4800 meters. Take shelter amongst rocks, under roots of trees, in hollow stumps or logs and quite often in the burrows of some other animals. Mainly nocturnal, although may hunt in day time also, hunts for rodents, pikas, birds and bird's eggs, sometimes reptiles and fish. Mostly present around open and forested areas in Khajjiar.

Conservation Status (CITES): Schedule III

Family: Ursidae

Mostly large, heavy animals strictly with plantigrade locomotion. They have a big head set with small eyes and rounded ears, a heavy body carried on thickest limbs and a small tail. Body covered with long and shaggy hairs under which small tail scarcely seen. Claws adopted for digging, being long and stout. Soles of feet usually devoid of hairs.

Genus: *Ursus* Linnaeus

1758. *Ursus* Linnaeus, *Syst. Nat., 10th ed.*, 1: 47

9. *Ursus thibetanus* G. [Baron] Cuvier (Black Bear)

(Plate 21 B)

1823. *Ursus thibetanus* G. [Baron] Cuvier, *Rech. Oss. Foss., Nouv. ed.*, 4: 325

Diagnostic Characters: A relatively large bear with black fur and shorter smoother and black claws. Body less clumsy and more compact as compared to other members of the family. Neck thick, head flattened, and forehead and muzzle in straight line, ears large, lower lip and chin white. It has a 'V' shaped white, yellow or buffy characteristic mark on breast sending up a branch on each side in front of the shoulder.

Distribution: Found in Kashmir, Himalayas and Assam extending eastward into China and Japan southward to Myanmar westward to Baluchistan roughly coincides with forest distribution in southern and eastern Asia.

Habits and Habitats: In summers, found near the limits of tree lines, between 3050 to 3660 metres but in winters comes down to 1525 metres. Lives in rock caves and hollows of the trees. Comes out at dusk for feeding. Omnivores and diet comprises more than 90% of plant materials. In some places the diet contains a sizeable portion of meat. Feed in plantations, where it may damage trees by stripping the bark and eating cambium, and in cultivated areas. This species can be observed in different habitat types especially around maize fields and higher riches of khajjiar.

Conservation Status (CITES): Schedule I

Family: Hynaeidae

In general appearance animals of this family resemble dog family but in their dentition and form of skull resemble those of Felidae. The powerful jaw of hyena and its large teeth adopted for bone crushing. The distinction in their teeth lies in the great size and strength of molars as compared with canines. Hyena has a broad head with large pointed and standing ears. Animal walks on its toes. The hind legs much bent, so that hind quarters of the body always lower than shoulders. Feet usually have four strong claws.

Genus: *Hyaena* Brisson

1762. *Hyaena* Brisson, *Regnum Animale, ed*. 2: 168

10. *Hyaena hyaena* Zimmermann (Striped Hyaena)

1780. *Hyaena straita* Zimmermann, *Georg. Gesch.* Ii, p. 256

Diagnostic Characters: Rather a dog like animal, the hind limbs shorter than the fore limbs. Head heavy with a shortened facial region, neck thick long and largely immobile. Eyes small but ears large and pointed sharply, set high on the head of animal. Tail short and feathery. Body covered with greyish brown fur which shows seasonal variations. Winter coat dirty brown-grey and has a well defined fur. Summers coat much shorter and coarser. Legs head and back have vertical stripes all over. Heavy dorsal crest of long hairs (mane) sharply defined from the rest of the coat distinguish the hyena. Body structures of males and females distinct. Females have three teats. Males have large pouch of naked skin located at the opening of anus.

Distribution: Quite common in many parts of India. Also found in southwest Asia and Africa.

Habits and Habitats: Inhabits grasslands and open country, especially where low hills and ravines offer holes and caves for shelter. A nocturnal animal, pairs usually go about together but sometimes group of five or six seen. Mainly a scavenger, feeds on dead animals or left over of animals killed by larger predators. Sometimes hunt smaller defenceless animals themselves.

Existence of this species in Khajjiar can be confirmed by indirect methods like pug marks and local residents including Himachal Forest Department officials but direct sighting is rare.

Family: Felidae

Members of this family, foremost of all carnivores. Head rounded and powerful. Jaws small and strong due to smallest number of molars, teeth particularly of cutting type, canines very large and sharp. Under surface of the feet have thick pads which make their footfall noiseless. Fur usually dense and short. Limbs of moderate length and very powerful. Vision adopted for night as well as day.

Genus: *Panthera* Oken

1816. *Panthera* Oken, *Lehrb. Naturgesch*, 3 (2): 1052

11. *Panthera pardus* (Linnaeus, 1758) (Leopard)

(Plate 21 D)

1758. *Felis pardus* Linnaeus, *Syst. Nat.*, 10th ed., 1: 41

Diagnostic Characters: Body large but slim and sleek with short hairs. Tail long and about two thirds of the length of head and body. Height up to 66 cm at the shoulder when standing. Colour of body varies from golden orange to pale greyish fawn, marked all over with small close set black rossets made of 4-5 concentric spots. Forehead and tail patterned with small spots.

Distribution: Indian leopard ranges over the whole country in suitable pockets up to 3000 m, widely distributed in Asia and Africa.

Habits and Habitat: Lives almost everywhere, not restricted only to dense forests or heavy covers but also thrives well in open country among rocks and scrubs. Primarily nocturnal but can hunt in day time being more tolerant to sun light. Kills and hunts anything that can be overpowered with safety, though mid-sized animals are preferred. Diet mostly consists of ungulates and monkeys, but rodents, reptiles, amphibians, insects, birds and fish also eaten.

Conservation Status (CITES): Schedule I

Order: Artiodactyla

Family: Cervidae

Deer family, have fine horns or osseous prominences called antlers, shed and renewed annually. Feet touch the ground only at the extremity of two principal toes, with two rudimentary toes (small hoofs) at the back of each foot. Eye pits or so called lachrymal sinuses constantly present. One or two glands covered with small tufts of hairs, on hind legs. Females have four teats.

Genus: ***Muntiacus*** **Rafinesque**

1815. *Muntiacus* Rafinesque, *Analyse de la Nature*: 56

12. *Muntiacus muntjac* (Zimmermann, 1780) (Barking Deer)

(Plate 21 E)

1780. *Cerves muntiacus* Zimmermann, *Georg. Gesch. Mensch. Vierf. Thiere.*, 2: 131

Diagnostic Characters: A small dear measuring 41-60 cm at shoulder. It has short and thin legs which gives stocky appearance. Colour of the fur glossy dark-chestnut with limbs internally, pubic region and tail beneath white, chin and lower jaw whitish, facial creases dark brown. Females smaller. Upper canines of the males well developed and used in self defence.

Distribution: Found throughout India except Jammu and Kashmir. Also found in Russia, central and eastern Asia, upper Myanmar and Java.

Habits and Habitat: Usually found in thickly wooded hills at the level of 1500 to 2450 metres. Found singly or in pairs grazing in outskirts of the forests or in open clearings. Diurnal in habits and feeds on leaves, grasses and wild fruits and also on succulent shoots. Carries its head and neck low, and hind quarters high; running action peculiar and not very elegant. This species is commonly encountered in Khajjiar area; easily identified by the call.

Family: Bovidae

Family of antelopes, goats, serows, gorals and cattle with horns composed of bony nucleus or core and a persistent bony sheath, present in both sexes or only in males. Pits usually present in all four feet, some of animals have inguinal pits. Eye pits usually present. Most of the animals have four mammas.

Genus: ***Naemorhedus*** **C.H. Smith**

1827. *Naemorhedus* C.H. Smith, In Griffith *et al.*, *Animal Kingdom*, 5: 532

13. *Naemorhedus sumatraensis* Bechstein, 1799 (Serow)

1799. *Naemorhedus sumatraensis* Bechstein, In Pennant, *Allgemeine ueber. Vierfuss. Thier*, 1: 98

Diagnostic Characters: A tall animal with large head, donkey like ears, thick neck and short sturdy fore and hind limbs. Conical and closely wrinkled horns present in both sexes. Tail short and busty. Body coat coarse but comparatively thin and colour may be from blackish-grey to red.

Distribution: In India found in the Himalayas from Jammu and Kashmir to Arunachal Pradesh. Also found in China, Indonesia, Malaysia, Myanmar and Thailand.

Habits and Habitat: Found at the elevation of 1850 to 3050 metres. Inhabit dense forest gorges of mountain streams where the massive boulders and caves provide shelter. Come out to feed in open slopes in morning and evening. More or less solitary creature but 4 to 5 individuals usually seen feeding on the same hill. Ascends to higher elevation in search of food in summers and returns to shelter of valleys in winters. This species is generally observed in oak-rhododendron forest during of Khajjiar.

Conservation Status (CITES): Schedule I

14. *Naemorhedus goral* (Hardwicke, 1825) (Goral)

(Plate 21 F)

1825. *Antelope goral* Hardwicke, *Trans. Linn. Soc. London,* 14: 518

Diagnostic Characters: A stout goat-like animal with arched back. Legs slender and light-tan in colour. A dark stripe extends down the spine and onto the forelegs. General colour of yellowish-gray. Chest and belly pale gray with a prominent white patch on the upper throat. Comparatively long tail covered with long black and gray hairs. Females paler than the males. Short pointed horns in both sexes, curved slightly backwards, with small irregular ridges.

Distribution: Found from Jammu and Kashmir, Himachal Pradesh and Punjab to Kumaon, and from Bhutan west to Nainital.

Habits and Habitat: Generally found at an elevation of 900 to 2750 metres. Prefers rocksy and precipitous slopes. Usually associate in small parties of five to six and can be seen grazing on scattered clumps of grasses. Active in morning and evening hours. In day time hide themselves or lie motionless, difficult to spot. Food consists of grasses, succulent roots and shoots. It is one of the common large mammals in the Khajjiar area.

Conservation Status (CITES): Schedule I

Order: Rodentia

Family: Sciuridae

Squirrels form a well marked group of elegant animals, mostly arboreal. Distinct post orbital process-most striking feature of the skull of the group; palate larger than other rodents. Molar teeth variously tubercled, some with blunt and others with sharp points. Tail typically long and bushy.

Genus: *Petaurista* Link

1795. *Petaurista* Link, *Zool. Bevtr.*, 1 (2): 52, 78

15. *Petaurista petaurista* Pallas, 1766 (Flying Squirrel)

1766. *Petaurista petaurista* Pallas, *Misc. Zool.*, p. 54

Diagnostic Characters: Presence of a membrane of skin between legs and arms, used to glide between trees. When at rest the parachute of membrane hardly noticeable. Leaps in the air with outspread limbs expand on the form of the parachute to its fullest extent. Eyes large and entire body has dark red colour. Nose, chin, eye-ring, feet and tail tip have black colour in addition to black spaces behind the ears.

Distribution: In India found in Assam, Bihar, Himachal Pradesh, Jammu and Kashmir, Meghalaya, Punjab, Uttaranchal, Uttar Pradesh and West Bengal. Outside India found in northern South Asia, southern China and southeast Asia.

Habits and Habitat: Found in moist evergreen broadleaf, temperate, coniferous and scrub forests, and in rocky areas as inland cliffs, mountain peaks between the elevation of 1818 to 3030 metres. Nocturnal and arboreal in habits, emerge from shelter at dusk and retire before the dawn. Usually roosts in the trees or shelter in branches. Food mainly consists of fruits and nuts, also leaves and branches, bark, gum and insects. This species is common in the area and also as a frequent visitor to forest rest house compound at Khajjiar.

Order: Rodentia

Family: Muridae

Members of this family, somewhat related to squirrels, and depicted as transition between them and rats. Small animals with soft fur-hairy and tufted tail. Live on tress. They have four molar teeth on each side, crown divided by closely folded lines of enamel. Lower incisors pointed. Forefeet usually have four toes and hind feet have five toes.

Genus: *Mus* Linnaeus

1758. *Mus* Linnaeus, *Syst. Nat., 10th ed.*, 1: 59

16. *Mus musculus* Linnaeus, 1758 (House Mouse)

(Plate 21 G)

1758. *Mus musculus* Linnaeus, *Syst. Nat.*, 10th ed., 1: 62

Diagnostic Characters: A small mammal measuring 7.5 to 10 cm from nose to the base of tail. Vary in colour from light brown to black; have short hair and a light belly. The ears and tail have little hair.

Distribution: In India found from Jammu and Kashmir to west Bengal, and northeast India. Elsewhere: Nepal, Myanmar, Thailand, Laos, Vietnam and China.

Habits and Habitat: Found in and around houses and commercial structures as well as in open fields and agricultural lands. Consume and contaminate food. In addition, often cause considerable agricultural and property damage. Presnt in human settlements, meadow and agriculture fields in Khajjiar area.

A Rhesus Monkey

B Black Bear

C Langur

D Leopard

E Barking Deer

F Goral

G House Mouse

Plate 21: Some mammals recorded in Khajjiar lake area

Table 5. 2: Systematic list of mammals and their status in Khajjiar lake area

Order – Insectivora **Family – Soricidae**	CITES	WPA(1971)	National Red Data
1. *Suncus murinus* (House Shrew)	-	-	-
Order – ChirpteraFamily – Vespertilionidae			
2. *Myotis muricola* (Nepalese Whiskered bat)	-	-	-
3. *Myotis blythii* (Lesser Mouse eared bat)	-	-	-
Order – PrimatesFamily – Cercopithecidae			
4. *Macaca mulatta* (Rhesus Monkey)	II	II	-
Family – Cercopithecidae			
5. Semnopithecus ajax (Hanuman Langur)	I	II	-
Order – CarnivoraFamily – Canidae			
6. *Vuples vulpes* (Himalayan Fox)	III	II	-
Family – Mustelidae			
7. *Martes flavigula* (Yellow throated marten)	III	II	-
8. *Mustela sibrica* (Himalayan Weasel)	III	II	-
Family – Ursidae			
9. *Ursus thibetanus* (Black Bear)	I	II	-
Family – Hynaeidae			
10. *Hyaena hyaena* (Striped Hyena)	-	III	-
Family – Felidae			
11. *Panthera pardus* (Leopard)	I	I	Vulnerable
Order – ArtiodactylaFamily – Cervidae			
12. *Muntiacus muntjac* (Barking deer)	-	III	-
Family – Bovidae			
13. *Nemarhedus sumatraensis* (Serow)	I	I	Vulnerable
14. *Nemarnhdus goral* (Goral)	I	III	-
Order – RodentiaFamily – Sciuridae			
15. *Petaurista petaurista* (Flying Squirrel)	-	II	-
Family – Muridae			
16. *Mus musculus* (House Mouse)	-	V	-

Table 5.3: Systematic list of different animal species present in Khajjiar area

S.n.	Order	Family	Scientific Name	Common Name
Annelida				
1	Oligochaeta	Lumbricidae	*Lumbricus castaneus*	Chestnut worm
Insecta				
2		Cordulegasteridae	*Antogaster basalis*	
3		Aeshnidae	***Anax immaculifrons***	Blue Darner
4		Libellulidae	*Orthetrum s. sabina*	Green Hawk Marsh
5	Odonata		*Orthetrum t. triangular*	Blue-tailed Forest Hawk
6			*Orthetrum pruinosum neglectum*	Crimson-tailed Marsh Hawk
7		***Palpopleura s. sexmaculata***	Blue-tailed Yellow Skimmer	
8			*Crocothemis servilia*	Ruddy Marsh Skimmer
9			*Trithemis festiva*	Black Stream Glider
10		Coenagrionidae	*Pseudagrionii* sp.	Marsh dart
11		Calopterygidae	*Neurobasis c. chinesis*	Stream Glory
12			*Ceracris n. nigricornis*	
13			*Truxalis indica*	
14			*Dnopherula (Aulacobothrus) decisus*	
15			*Aiolopus thalassinus tamulus*	
16	Orthoptera	Acrididae	*Locusta migratoria danica*	
17			*Pseudosphingonotus savignyi*	
18			*Spathosternum p. prasiniferum*	
19			*Oxya fuscovittata*	
20			*Eucoptacra saturata*	

Contd.

Table 5.3: Contd....

S.n.	Order	Family	Scientific Name	Common Name
21			*Eyprepocnemis rosea*	
22			*Tylotropidius varicornis*	
23			*Diabolocatantops innotabilis*	
23			*Xenocatantops karnyi*	
25			*Chondracris rosea*	
26			*Cyrtacanthacris tatarica*	
27			*Patanga succineta*	
28		Pyrgomorphidae	*Aularches punctatus*	
29			*Poikilocerus pictus*	
30			*Eucriotettix grandis*	
31			*Teredorus frontails*	
32		Tetrigidae	*Coptotettix consperus*	
33			*Ergatettix dorsiferus*	
34			*Ergatettix guentheri*	
35			*Hedotettix costatus*	
36		Gryllidae	*Acheta domesticus*	
37			*Gryllus bimaculatus*	
38			*Loxoblemmus equestris*	
39			*Teleogryllus occipitalis*	
40		Conocephalidae	*Conocephalus maculatus*	
41		Pentatomidae	*Dalpada nigricollis*	Shield bug
42			*Erthesina fulla*	Sting bug
43	Hempitera		*Prionolomia cardoni*	
44		Largidae	*Physopelta gutta*	
45			*Physopelta schlanbuschi*	

Contd.

Table 5.3: Contd....

S.n.	Order	Family	Scientific Name	Common Name
46	Homoptera	Cicadiae	*Platylomia saturata*	Cicada
47	Coleoptera	Carabidae	*Carabus boysi*	
48			*Calosoma* sp.	
49		Dytiscidae	*Dytiscus punctalis*	
50		Scarabaeidae	*Clinteria* sp.	
51			*Mimela pictoralis*	
52			*Onitis subopacus*	
53			*Popilla* sp.	
54			*Melolontha furcicauda*	
55			*Anomala rufiventris*	
56		Hydrophilidae	*Helochaes* sp.	
57		Coccinellidae	*Coccinella septempunctata*	Lady bird beetle
58		Meloidae	*Myllabris pustulata*	
59		Chrysomellidae	*Cleorina* sp.	
60		Cerambycidae	*Dorysthenes (Lophosternus) huegeli*	
61			*Macrotoma* sp.	
62		Papilionidae	*Papilio protenor*	The Spangle
63	Lepidoptera		*Papilio polyctor polyctor*	The Common peacock
64			*Parnassius hardkwickei hardwickei*	The Common Blue Apollo
65		Pieridae	*Delias belladonna horsfieldi*	The Hill jezebel

Contd.

Table 5.3: Contd....

S.n.	Order	Family	Scientific Name	Common Name
66			*Pieris canidia indica*	The Indian Cabbage White
67			*Catopsillia crocale* Emigrant	The Common
68			*Gonepteryx rhamni nepalensis*	The Common Brimstone
69			*Eurema hecabe fimbriata*	The Common Grass Yellow
70			*Colias electo fieldi*	The Dark Clouded Yellow
71		Danaidae	*Danaus genutia*	The Common Tiger
72			*Parantica sita sita*	The Chestnut Tiger
73		Satyridae	*Mycalesis perseus blasius*	The Common Bushbrown
74			*Lethe insana insana*	The Common Forester
75			*Lethe scanda*	The Blue Forester
76			*Lethe verma verma* treebrown	The Straight-Banded
77			*Lasiommata schakra schakra*	The Common Wall

Contd.

Table 5.3: Contd....

S.n.	Order	Family	Scientific Name	Common Name
78			*Aulocera swaha swaha*	The Common Satyr
79			*Aulocera saraswati*	The Striated Satyr
80			*Callerebia annade*	The Ringed Argus
81			*Ypthima nareda nareda*	The Large Threering
82			*Ypthima ceylonica hubneri*	The Common Fourring
83			*Ypthima sakra nikaea*	The Himalayan Fivering
84			*Melanitis leda ismene*	The Common Evening Brown
85			*Athyma opalina*	The Himalayan Sergeant
86			*Athyma asura*	The Studded Sergeant
87			*Neptis mahendra*	The Himalayan Sailer
88			*Neptis hylas astola*	The Common Sailer
89		Nymphalidae	*Pseudergolis wedah*	The Tabby
90			*Precis iphita*	The Chocolate pansy
91			*Cynthia cardui*	The Painted Lady

Contd.

Table 5.3: Contd....

S.n.	**Order**	**Family**	**Scientific Name**	**Common Name**
92			*Vanessa indica*	The indian Red Admiral
93			*Kaniska canace*	The Blue Admiral
94			*Aglais cashmirensis*	The Indian Toroiseshell
95			*Childrena childreni*	The Large Silverstripe
96			*Issoria lathonia*	The Queen of Spain Fritillary
97		Acraeidae	*Acraea issoria anomala*	The Yellow Coster
98			*Libythea myrrha*	The Club
99		Erycinidae	*Libythea lepita*	The Common Beak
100			*Dodona durga*	The Common Punch
101			*Pseudozizeeria maha*	The Pale Grass Blue
102			*Lampides boeticus*	The Peablue
103		Lycaenidae	*Lycaena pavana*	The White-Bordered
			Copper	
104			*Heliophorus sena*	The Sorrel Sapphire
105			*Castalius rosimon*	The Common Pierrot

Contd.

Table 5.3: Contd....

S.n.	Order	Family	Scientific Name	Common Name
106			*Ropala manea schistecea*	The state Flash
107		Hesperiidae	*Coladenia dan*	The Fulvous Pied Flat
108			*Sarangesa purendra*	The Spotted Small Flat
109			*Polytrema eltola*	The Yellow-Spot Swift
110			*Borbo bevani*	The Bevan's Swift
111		Asilidae	*Maschimus* sp.	
112	Diptera	Calliphoridae	*Chrysomya rufifacies*	
113			*Calliphora vomitoria*	
114			*Phaenicia Sericata*	
115		Muscidae	*Musca domestica*	House fly
116			*Melecta himalayana*	Cuckoo bee
117			*Anthophora zonata*	Flower bee
118	Hymenoptera	Apidae	*Bombus trifasciatus*	Bumble bee
119			*Apis dorsata*	Honeybee
120			*Apis cerana indica*	Desi makhi
121		Vespidae	*Polistes hebrsea*	Yellow wasp
122			*Vespa basalis*	Social wasp
Mollusca				
123	—	Not identified	Not identified	Slug
Fish				

Contd.

Table 5.3: Contd....

S.n.	Order	Family	Scientific Name	Common Name
124	Cypriniformes	Cyprinidae	*Cyprinus carpio*	Common Carp
Amphibians				
125	Anura	**Bufonidae**	*Bufo himalayanus*	Toad
126		Ranidae	*Rana liebigii*	Frog
Reptiles				
127	Squamata	Agamidae	*Laudakia tuberculata*	Agama Lizard
128	Scincidae	***Scincella himalayanus***		Himalayan Ground Skink
129	Colubridae	*Amphiesma platyceps*	Eastern Keelback	
130		*Ptyas mucosus*	Indian Rat Snake	
Aves				
131	Ciconiformes	Ardeidae	*Bubulcus ibis*	Cattle Egret
132			*Milvus migrans*	Black Kite
133			*Gypaetus barbatus*	Lammergeir
134	Falconiformes	Accipitridae	*Neophron percnopterus*	Egyptian Vulture
135			*Gyps himalayensis*	Himalayan Griffon
136			*Aegypius monachus*	Cinereous Vulture
137			*Ictinaetus malayensis*	Black Eagle
138	Galliformes	Phasianidae	*Pucrasia*	Koklass *macrolopha* Phesant
139		*Lophaphorus impejanus*	Himalayan Monal	

Contd.

Table 5.3: Contd....

S.n.	Order	Family	Scientific Name	Common Name
140			*Lophura leucomelanos*	Kalij Pheasant
141	Galliformes	Recurvirostridae	*Himantopus himantopus*	Black-winged Stilt
142			*Columba livia*	Blue Rock Pigeon
143	Columbiformes	Columbidae	*Streptopeliaorientalis*	Oriental Turtle Dove
144			*Treron sphenura*	Wedge-tailed Green Pigeon
145	Psittaciformes	Psittacidae	*Psittacula himalayana*	Slaty Headed Parakeet
146			*Psittacula cynocephala*	Plum Headed Parakeet
147	Cuculiformes	Cuculidae	*Cuculus canorus*	Common Cuckoo
148	Strigiformes	Strigidae	*Glaucidium cuculoids*	Asian Barred Owlet
149	Caprimuls giforme	Caprimulgidae	*Caprimulgus macrurus*	Large-tailed Nightjar
150	Apodiformis	Apodidae	*Apus affinis*	House (little) Swift
151	Coraciiformes	Alcedinidae	*Halcyon smyrnensis*	White-breasted Kingfisher152
		Megaceryle lugubris Kingfisher	Greater Pied	
153		Upupidae	*Upupa epops*	Common Hoopoe
154		Capitonidae	*megalaima virens*	Great Barbet

Contd.

Table 5.3: Contd....

S.n.	Order	Family	Scientific Name	Common Name
155	Piciformes	Picidae	*Dendrocopos himalayensis*	Himalayan Woodpecker
156			*Picus squamatus*	Scaly-bellied Woodpecker
157			*Picus canus*	Grey-headed Woodpecker
158		Hirundinidae	*Hirundo daurica*	Red-rumped Swallow
159		Motacillidae	*Motacilla maderaspatensis*	Large Pied Wagtail
160		Campephagidae	*Pericrocotus ethologus*	Long-tailed Minivet
161		Pycnonotidae	*Pycnonotus leucogenys*	Himalayan Bulbul
162	Passeriformes		*Hypsipetes leucocephalus*	Black bulbul
163		Laniidae	*Lanius vittatus*	Bay-backed Shrike
164		Muscicapidae	*Monticola cinclorhynchus*	Blue-capped Rock Thrush
165		*Monticola rufiventris*	Chestnut-belliled Rock Thrush	
166		*Myopphonus caeruleus*	Blue Whistling Thrush	
167		*Turdus albocintus*	White-collared Blackbird	
168		*Turdus boulboul*	Grey-winged Blackbird	
169			*Turdus viscivorus*	Mistle Thrush
170		*Chaimarrornis leucocephalus*		White-capped Water Redstar
171		*Rhyacornis fuliginosus*	Plumbeous Water	

Con

Table 5.3: Contd....

S.n.	Order	Family	Scientific Name	Common Name
			Redstart	
172		*Enicurus maculatus*	Spotted Forktail	
173		*Saxicola torquata*	Common Stonechat	
174			*Saxicola ferrea*	Grey Bushchat
175		*Garrulus lineatus*	Streaked Laughingthrush	
176		*Garrulus variegatus*	Variegated Laughingthrush	
177			*Heterophasia capistrata*	Rufous Sibia
178		*Phylloscopus collybita*	Common Chiffchaff	
179		*Phylloscopus trochiloids*	Greenish Warbler	
180		*Seicercus burkii*	Golden-spectacled Warbler	
181		*Muscicapa sibirica*	Dark-sided Flycatcher	
182		*Ficedula wastermanni*	Little Pied Flycatcher	
183		*Ficedula superciliaris*	Ultramarine Flycatcher	
184		*Eumyias thalassina*	Verditer Flycatcher	
185		*Niltava sundara*	Rufous-bellied Niltava	
186		Aegithalidae	*Aegithalos concinnus*	Red-headed
Tit				
187		Paridae	*Parus melanolophus*	Spot-winged Tit
188			*Parus major*	Great Tit
189		*Parus monticolus*	Green-backed Tit	
190		Sittidae	*Sitta leucopsis*	White-cheeked Nuthatch
191		Certhiidae	*Certhia himalayana*	Bar-tailed Treecreeper

Contd.

Table 5.3: Contd....

S.n.	Order	Family	Scientific Name	Common Name
192		Zosteropidae	*Zosterops palpebrosus*	Oriental White-eye
193		Emberizidae	*Melophus lathami*	Crested Bunting
194			*Emberiza cia*	Rock Bunting
195		Fringillidae	*Carduelis spinoides*	Yellow-breasted Greenfinch
196			*Mycerobas icterioides* Grosbeak	Black-and-yellow
197		Passeridae	*Passer domesticus*	House Sparrow
198			*Passer rutilans*	Russet Sparrow
199		Sturnidae	*Acridotheres tristis*	Common Myna
200			*Acridotheres fuscus*	Jungle Myna
201		Dicruridae	*Dicrurus macrocerus*	Black Drango
202		Corvidae	*Garrulus glandarius*	Eurasian jay
203			*Garrulus lanceolatus*	Black Headed Jay
04			*Urocissa flavirostis*	yellow-billed Blue Magpie
205			*Urocissa erythrorhyncha*	Red billed blue Magpie
206			*Dendrocitta formasae*	Grey Treepie

Contd.

Table 5.3: Contd....

S.n.	Order	Family	Scientific Name	Common Name
207			*Corvus macrorhynchos*	Large-billed Crow
				Mammal
208	Insectivora	Soricidae	*Suncus murinus*	House Shrew
209	Chiroptera	Vespertilionidae	*Myotis muricola*	Nepalese whiskered bat
210			*Myotis blythii*	LesserMouse-eared bat
211	Primates	Cercopithecidae	*Macaca mulatta*	Rhesus Monkey
212		*Semnopithecus ajax*	Hanuman Langur	
213	Carnivora	Canidae	*Vulpes vulpes*	Himalayan Fox
214		Mustelidae	*Martes flavigula*	Yellow throated Martin
215			*Mustela sibrica*	Himalayan Weasel
216		Ursidae	*Ursus thibetanus*	Black Bear
217		Hynaeidae	*Hyaena hyaena*	Striped Hyaena
218		Felidae	*Panthera pardus*	Leopard
219	Artiodactyla	Cervidae	*Muntiacus muntjac*	Barking Deer
220		Bovidae	*Nemarhedus sumatraensis*	Serow
221			*Nemarnhdus goral*	Goral
222	Rodentia	Sciuridae	*Petaurista petaurista*	Flying Squirrel
223		Muridae	*Mus musculus*	House mouse

Chapter 6

Biodiversity of Khajjiar: Threats and Conservation

Relation of man with bio-diversity is as old as the evolution of man itself. He has been dependent upon the biodiversity for fulfillment of his entire livelihood needs. The ancient Indian literature is replete with reference to animal and plant life affecting the lives of human beings on day to day basis. Many of the plants and animals have been domesticated over the period of time. Man has also discovered curative uses of plants and animals found in his immediate vicinity, and have associated religious value to the most important of such plants and animals. But degradation of the Indian Himalaya, of which the state of Himachal Pradesh is a part, is having a profound influence on all ecosystems of the region resulting in the loss of biodiversity. It is, therefore, very important to conserve the ecology and biodiversity of this fast deteriorating fragile ecosystem. Several endemic species have evolved in this area and are particularly adapted to the Himalayan way of life, especially at high altitudes above timber line. There are also many valuable animal resources from scientific and aesthetic points of view, and several of these are being threatened due to large scale destruction of their habitats. Many interesting unknown taxa may become extinct before they are even discovered. Considering the ecosystem as a gene bank, if populations of species are reduced, there may be a genetic drift and gene loss which is potentially an important loss.

Extinction rates which are far higher than the normal background rates are resulting in rapid loss of biodiversity of animals. By deprived of species we are losing potential contributors to future food, medicine and valuable links in natural and biological cycles. Although extinction is a natural phenomenon, fossil records show that on an average one species dies out every 100 years. During the last 200 years the rate of extinction has been at least 40 times greater than this. The main causes of extinction are habitat loss and habitat degradation. Changes in land use patterns had a detrimental impact on habitats, which have been fragmented and reduced in extent and diversity (Birdlife International, 2001).

Therefore, present work, in the long run, will help in evolving the conservation measures to protect and assure the survival of the threatened and endangered species, in addition to total improvement of ecological efficiency of the area. Secondly, different groups of animals are very good bio-indicators of any disturbance in the ecosystem, so, present record on diversity, status etc. of faunal elements shall be of great use in monitoring changes in the Khajjiar area, especially due to habitat degradation, deforestation, organic and inorganic pollution. This work will also be very useful in terms of bio-diversity conservation, as, an effective way to conserve the bio-diversity is to save the places where they live, feed and breed. Therefore, present work in the long run will play a very significant role in providing a reliable database upon which to base effective conservation measures for fauna of the Himalayan region.

Like India, the faunal and floral diversity in Himachal Pradesh is also very rich and diversified, primarily due to varied climatic conditions ranging from tropical in the foothills to arctic environment in the Trans-Himalayan region. Moreover, historical influx of fauna from adjacent bio-geographical regions and subsequent speciation in relation to local environment has greatly enriched the animal resources of the area. There is a pronounced dominance of Palaearctic and endemic animals above timber line (3000 m), and largely Oriental and some Palaearctic and Ethiopian elements at lower and middle altitudes. Thus, rich biodiversity of Himachal Pradesh has sustained population and hill communities from times immemorial (Mehta and Julka, 2002). Similarly Khajjiar area is also having rich diversity. Observations of 223 of faunal species from Khajjiar area showed that the alpine meadow of Khajjiar supported a good diversity of different animal groups. The Khajjiar area supports more than 8.7% of the total fauna of the state as Mehta (2005) has documented 2,542 animal species in varied ecosystems of the state of Himachal Pradesh.

Though most invertebrate species are cosmopolitan, others can be quite specific in the habitat requirements. The quantitative concentration of the invertebrates varies from season to season and also from one to other regions of the country. However, in India, maximum concentration of all the species has been recorded in the monsoon and post monsoon seasons (June to September) when the growth of all types of vegetation in the natural grassfields as well as in the agricultural fields is in optimum conditions. The comparison of number of species of invertebrates from the area and known from Himachal Pradesh (Mehta, 2005) revealed the presence of 6.6% of the invertebrate fauna of the state in the Khajjiar area.

Himachal Pradesh has a small geographical area of 55,673 square kilometres which is only 1.7% of the country but it harbours more than 7% of the total fauna of India. Invertebrates constitute 88.4% and vertebrates 11.6% of the Himachal fauna. Insects and other arthropods form a predominant group (4,641 species) among invertebrates, whereas vertebrates are dominated by birds comprising about 447 (610, revised) species (Mehta and Julka, 2002). Similarly, in khajjiar area Invertebrates constitute more than 55% of the total fauna (Fig. 7). Insects form around 55% of the total fauna of the study area and 98% of the invertebrates (Fig. 8).

Orthoptera is one of the largest insect orders with over 20,000 species known to the science throughout the world (Gillot, 2005). More than 1,750 species, about 10% of the total world species, have been recorded from India (Alfred and Ramakrishna,

2004). Khajjiar area supports a total of 29 species of Orthoptera. This represents more than 17% of the state fauna and form more than 23% of the total invertebrate fauna of the Khajjiar area (Fig. 9). If we analyse the Orthopterans of the state number of species in each family showed a decreasing trend with increase in altitude from Shivalik to the Greater Himalayan zone, and diversity decreased from 94 species in Shivalik zone to 26 in the Greater Himalayan, through 58 species in the Middle Himalayan zone. Moreover, there is a correlation of abundance of grasshopper species with the plant communities on which they depend which in turn has been linked to the climatic, physiographic, edaphic and biotic factors which control the plant distribution, therefore, presence of 17% of the total orthoptera fauna of the state in the Khajjiar area situated in middle Himalayan zone is justified.

Only 6% world Dipteran diversity has been reported from India which is 6,093 species under 1,075 genera and 87 families (Datta, 1998). 5 species of dipterans are present which constitute around 4% of the invertebrate (Fig. 9) and 1.7% of the total fauna of Khajjiar area (Fig. 10). The faunal diversity of Khajjiar shows the presece of 7 species of Hymenoptera forming around 6% of the invertebrate (Fig. 9) and 3% of the total fauna of Khajjiar area of Himachal Pradesh (Fig. 10). Almost equally, India holds about 8.3% of total world's hymenoptera with about 10,000 species (Jonathan, 1998).

The World List of Four Vertebrate Groups, mammals, birds, reptiles and amphibians, comprises 27,298 species, consisting of 4,809 mammals, 9,881 birds, 7,828 reptiles and 4,780 amphibians (Sibley and Monroe, 1990; Nowak, 1999; Myers *et al.*, 2000). India, a mega diverse country with only 2.4% of world's land area, accounts for 7-8% of the recorded species of the world, including 45,500 species of plants and 91,000 species of animals. Likewise a total of 100 species representing around 45% of the total fauna are there in the Khajjiar lake area (Figs. 1 and 7).

Laudakia tuberculata is reported for the first time from Chamba district (Singh and Banyal, 2014). This is a low risk least concerned species according to Conservation Assessment and Management Plan (CAMP). This species has not yet been assessed for IUCN red list (IUCN 2013). But due to heavy intervention of human activities in the Khajjiar area, ecology and habitat of this species is disturbed. There is a heavy load of tourism which is resulting in construction activities, hence, reduction and disturbance in wild habitat (Singh and Banyal, 2012). Earlier, Saikia et al. (2007) have given its general report from the state but none of the contribution reported its presence in Chamba district. 77 species of birds present in Khajjiar area represents 77% of the vertebrate and 34.5% of the total fauna of Khajjiar (Figs. 11 and 10). This data supports the earlier findings of Ali (1949) who enlisted about 225 species of birds from different regions of Western Himalaya.

The Mammalian fauna of Himachal Pradesh is an admixture of Palaearctic and oriental elements since the state lies in transition zone of two biogeographical realms. Brown Bear, Lynx, Alpine Weasel, Mountain Noctule etc. are some of Palaearctic representatives of State which probably reached from the Hindukush Mountain and Russian Uzbekistan. Some of the representatives of Oriental fauna of the State include Leopard Cat, Yellow throat Marten, Himalayan Palm Civet, Indian Pangolin, Grey Goral, Barking Deer, Bandicoot Rat, Bush Rat, Flying Fox, False Vampire, Fulvc Leaf-nosed Bat, Musk Shrew etc. (Roberts, 1977).

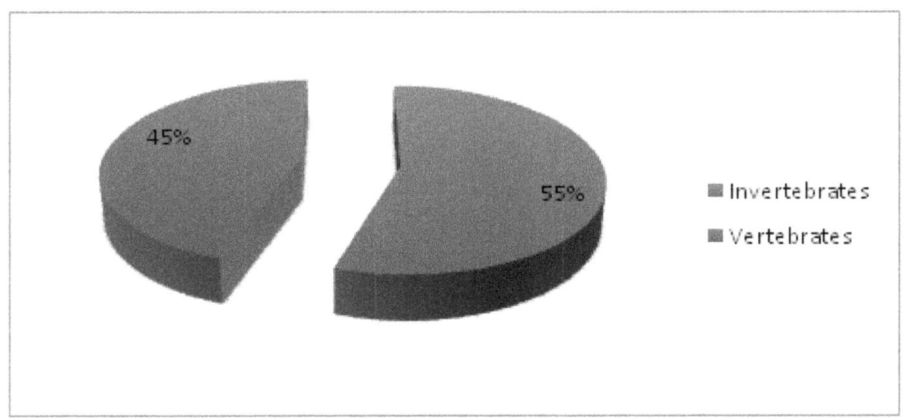

Figure 6.1: Percentage of invertebrates and vertebrates fauna in Khajjiar lake area

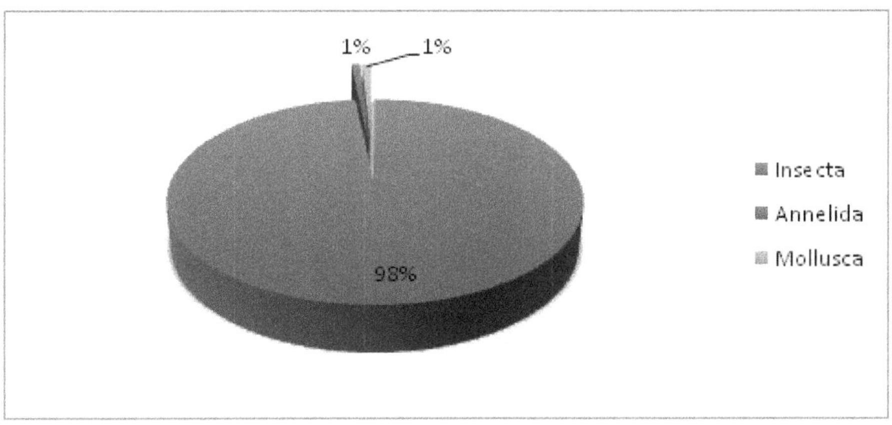

Figure 6.2: Percentage of different invertebrate groups in Khajjiar area

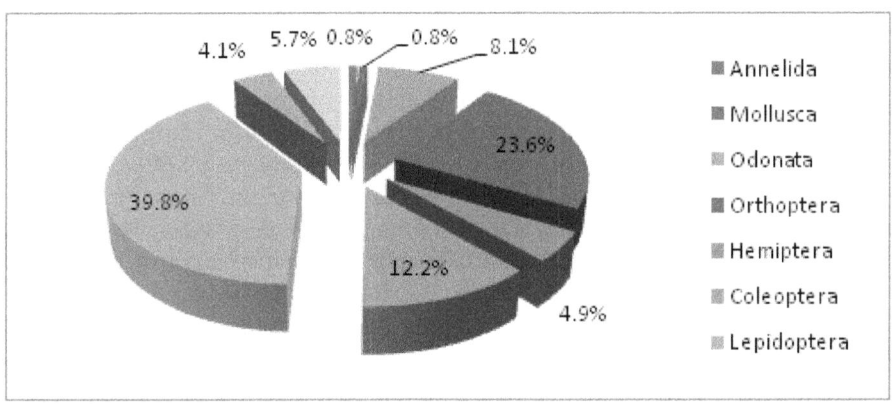

Figure6.3: Percentage of different invertebrate groups present in Khajjiar area

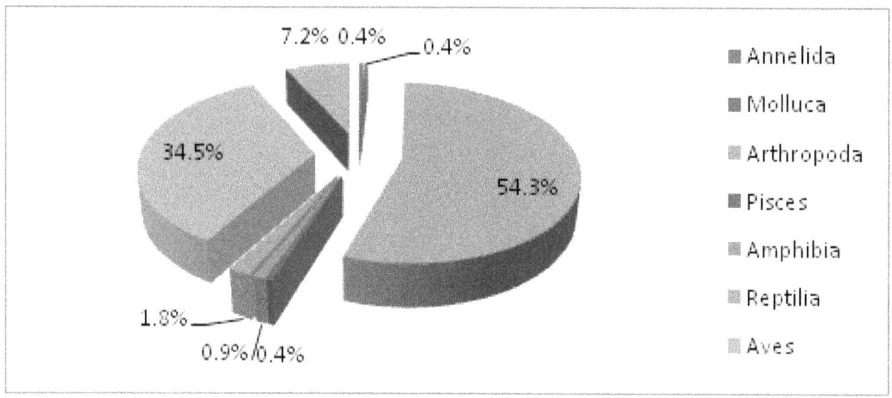

Figure 6.4:: Percentage of different invertebrate and vertebrate groups in Khajjiar area

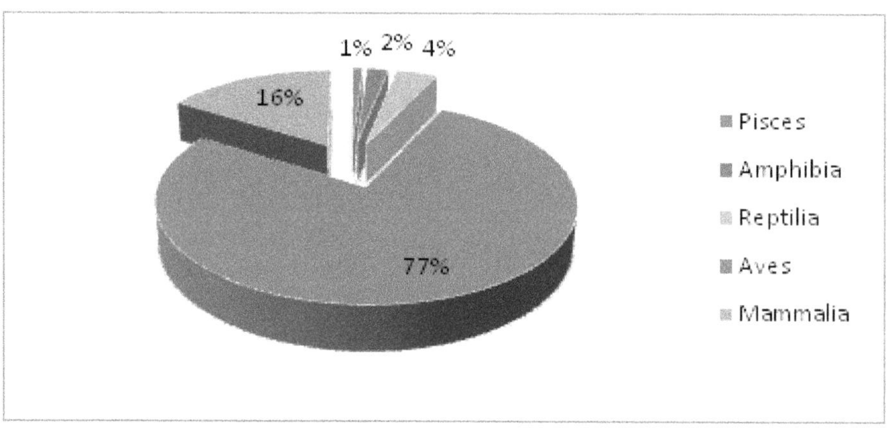

Figure 6.5: Percentage of vertebrate groups in Khajjiar area

In Khajjiar, 16 species of mammals are present, which belong to 14 genera, 12 families and 6 orders from the Khajjiar area . Of the 16 species of mammals, 11 species belong to the category of large mammals and 5 species to the category of small mammals. Nine species are listed as threatened in Convention in Trade of Endangered Species (CITES) under different schedules. Five species namely *Semnopithecus ajax, Ursus thibetanus, Panthera pardus, Naemorhedus sumatraensis*and *Naemorhedus goral* are placed in schedule I, *Macaca mulatta* in Schedule II and *Vulpes vulpes, Martes flavigula* and *Mustela sibrica* under schedule III. Out of a total of 16 species, 13 are placed under the Indian Wildlife Protection Act 1972. Two species *Panthera pardus* and *Naemorhedus sumatraensis*are kept under schedule I. Same species are considered as vulnerable species according to National Red Data.

Mammals of the study area have different feeding habits. Animals like Fox, Marten, Weasel, Hyena and Leopard are typical carnivorous and prey upon other small animals or eggs of birds. Animals like Black Bear are omnivores and its diet comprises more than 90% of plant materials (Hwang and Garshelis, 2007). Crops of locals are invaded by Black Bear. They feed on plantations, where they damage trees by stripping the bark and eating cambium, and in cultivated areas (Mizukami *et al.*, 2005, Gong and Harris 2006, Vinitpornsawan *et al.*, 2006). In some places the diet contains a sizeable portion of meat (Hwang *et al.*, 2002). Other animals like Rhesus Monkeys, Hanuman Langurs, Barking Deer and Goral are primarily herbivorous, generally feed on leaves, fruits, buds and flowers but sometimes also feed on insects, tree bark and gum (Jerdan, 1984). Feeding habits have been modified with availability of food and human development in the area. Leopard which is essentially a carnivorous animal, feeds on monkeys and ungulates in the normal wild state, comes out of the forest and lives near the human settlements. Leopards are forced to come out of their habitation in search of monkeys which is never safe for this large animal in view of encounters with human. On many occasions Leopards attack the domestic animals of local inhabitants. Similarly, during shortage of food in the forest, there is very less leftover for hyenas to eat which compel this animal to search for alternative source of food.

Ecological equilibrium of the study area is no more in a balanced stage due to developments and human interventions. With the passage of time, natural food plants of Monkeys and Langurs have decreased in the forest and these animals have come out of their natural habitat and forced to live near or around human population. Similar observations have been made in some studies conducted in different parts of country in recent past. Southwick and Siddiqi (1994) reported that in the northern parts of our country normally 86% of rhesus monkey population depends entirely upon human settlements for their food; however, only 14.4% of the rhesus macaques live in isolation from humans and do not rely on them at all for food. Leopard prefers easily available food in the form of domestic animals and stray cattle, therefore, natural check on monkey is not there. Feeding habits of monkeys and langurs have also changed. Now they havebecome more dependent on human left over like baked or cooked food available near the human population, offered by tourists and leftover of the hotels. Rhesus and Langurs usually raid the crops of the natives and cause huge economic loss to them. A new kind of conflict is developed between the ecology of these animals and local farmers. Various incidences of violence of monkeys against tourists are commonly noticed (Singh and Banyal, 2012).

Similarly, with increasing intervention of man into forest incidences of encounters between man and bear have also increased in the Kalatop-Khajjiar. Most of the places are remote and there are no access to the vehicle so local people have been using the forest path and sometimes it gives rise to bear-human interface. With decreasing food resources, bears are forced to raid maize and other crops of farmers. Amount of the destruction of the crop is much higher than they actually eat. Earlier when the food resources were available in the forest this kind of raids of crops were rarely observed as informed by people. Although Monkeys, Langurs and Bears are in conflict with humans but no incidence of their killing was recorded from the area.

Another important concern of ecology is of stray cattle and cows. These are in huge numbers and can be seen grazing in and around the Khajjiar lake. Further, the number of these stray animals is increasing day by day. This leads to increased addition of faecal matter in the lake which is leading to eutrophication of the lake. Many times these stray animals enter forest for grazing and destroy undergrowth of forest. With ever increasing number of tourists reaching Khajjiar every year, the number of hotels in the area is also multiplying. This is good for general socio- economic development of the area but has adverse impacts on ecological health. Many tourists visit deep in the forests and enjoy tracking in the hills. Hotels and tourists produce a large quantity of non degradable garbage which accumulates in and around the lake and also deep into the forest. This non degradable litter also interferes with the rejuvenation of forest organic mass which impacts the floral and faunal diversity.

Biodiversity has three important categories of values viz., productive use value, consumptive use value and indirect values. Productive use value is a value assigned to products that are commercially harvested for exchange in formal markets and is, therefore, the only value of biological resources that is reflected in national income accounts. On the other hand, consumptive use value is the value placed on natural products that are consumed directly. The value of such goods can be considerable. Indirect values are related primarily with the functioning of ecosystems, do not normally appear in national accounting systems, but they may far outweigh the consumptive and non-consumptive values (Alfred *et al.*, 1998). The bio-diversity in the Khajjiar lake area, like other parts of Himachal Pradesh, is very rich and diversified, but, in recent years, the present Khajjiar in particular and the state in general have come under a strong threshold of development. Natural ecosystems/habitats have been over-exploited and even destroyed by the rapidly increasing human population and tourist inflow. A number of endemic and restricted range species found in the area/region are facing threat to their existence (Vedwan and Rhodes, 2001).

There already exist many programmes/acts which if strictly followed, can play an important role in the preservation of bio-diversity of not only urbanized areas but of rural and forested areas too. The Wildlife Protection Act (1972), the National Wildlife Action Plan (1983), National Biodiversity Strategy and Action Plan (NBSAP) etc., all envisage objectives which aim at all active protection and development of forest resources, conserving nation's biodiversity and strengthening efforts to protect wild species and varieties. Convention on Biological Diversity (1992) emphasizes various objectives all of which have one thing in common i.e., protection of nature and natural resources. Moreover, India has a well developed Protected Area Network (PA's) comprising 89 National Parks (covering an area of 37,530.76 km^2, or 1.14% of the country's geographical area) and 489 Wildlife Sanctuaries (1,17,042.04 Km^2, or 3.56% of the country's geographical area). The KhajjiarLake is also a part of the richest and oldest preserved wildlife sanctuaries of the country. Put together the 578 PAs cover about 4.70% of the country's geographical area (Rodgers *et al.*, 2002).

The Khajjiar area of Himachal Pradesh has seen a tremendous increase in population in the last decade, due to which natural habitats are in great pressure. The future of the unique Himalayan Wildlife found in the area of Khajjiar, therefore, requires immediate involvement of scientific inputs, political will and collective public participation in saving bio-diversity from imminent danger of appalling extinction. Keeping in view the imminent dangers of extinction to bio-diversity following measures are proposed:

1. Keeping in view the eutrophication of the lake, entry of stray and domesticated animals to the Khajjiar meadow should be immediately checked.
2. A barbet wire fencing of the lake and meadow will prevent stray animals to the lake.
3. Proper management of horse dung is the need of the hour.
4. Proper management of non-biodegradable as well biodegradable waste like plastics and bottles is required.
5. There should be some check on the entry of tourists keeping in view the huge population of tourists visiting the meadow every year.
6. Proper management of waste material from hotels is urgently required.
7. Proper guidelines for tourism sports in the Khajjiar meadow should be prepared.
8. Keeping in view the disturbance created to the wildlife by vehicles in the form of horn, air pollution, noise pollution etc., traffic laws should be strictly implemented on all the vehicles entering the sanctuary area.
9. Entry of tourists into the forest patches should be restricted.
10. Multiplication and breeding of threatened species of fauna through modern techniques of tissue culture and biotechnology be encouraged.
11. To establish conservation pockets for the rare and endemic faunal species found in the Khajjiar area.
12. Documentation of local resources and support for threatened species.
13. Inventorisation and monitoring of processes adversely impacting bio-diversity.
14. Development of *in-situ* measures for bio-diversity conservation, as a complement to *ex-situ* approaches.
15. Restoration of degraded habitats and recovery of endangered species.
16. Strengthen ongoing conservation measures on bio-diversity including conservation of habitats for wildlife.
17. Rehabilitation of rural poor recorded in the protected area of Khajjiar.
18. Adopting economically and socially sound measures that act as incentives for conservation and sustainable use of components of biodiversity.
19. Promoting scientific and technical co-operation among the Himalayan researchers.
20. To undertake these programmes among rural people, farmers and shepherds.
21. Developing educational and public awareness programmes with respect to conservation and sustainable use of biodiversity.

Chapter 7

Executive Summary

Biodiversity is our natural wealth. Its conservation is important for both economic and ethical reasons. It provides us goods and services fundamental to our survival, including clean air, fresh water, medicines and shelter. It enables us to adapt to changing needs and circumstances. For example, forested ecosystems provide us fuels, medicines, construction material and wildlife habitat; wetlands and riparian areas protect water quality and aquatic life; oceans provide food and regulate climate; and agro-ecosystems produce food. Biodiversity also provides people with recreational, psychological, emotional and spiritual enjoyment. Some people are of the opinion that we should protect and restore biodiversity because of its benefits to mankind, while others believe that it is our moral obligation to care about biodiversity simply because all species have the right to live and to be valued in nature, whether we understand their benefits to human beings or not.

Khajjiar area situated at 32° 26′ north and 76° 32′east, at an altitude of about 6,300 feet (1,920 meters) above sea level between Chamba and Dalhousie is one of the most favourable tourist destinations of Himachal Pradesh. Therefore, this area has seen a tremendous increase in population in the last decade, due to which natural habitats are in great pressure. The future of the unique Himalayan wildlife found in the Khajjiar, therefore, requires immediate involvement of scientific inputs, political will and collective public participation in saving bio-diversity from impending danger of appalling extinction.

Khajjiar area of Himachal Pradesh harbours 223 species of different faunal groups corresponding to around 8.7% of the total fauna of the state, spread over 193 genera, 79 families and 32 orders. Analysis of this diversity shows that class Aves dominate the fauna of Khajjiar with 77 species, followed by Lepidoptera (49 species), Orthoptera (29), Mammalia (16), Coleoptera (15), Odonata (10), Hymenoptera (7), Hemiptera and Diptera (5 each), Reptilia (4) and Amphibia (2 species). It is further explored that Mollusca, Oligochaeta, Homoptera and Pisces are least represented groups of fauna with a single species each in Khajjiar area.

Currently there is a presence of 123 species of invertebrates belonging to 110 genera, spread over 30 families and 10 orders from Khajjiar area. Of these, Lepidoptera (49 species) is the most dominant invertebrate order in the area, followed by Orthoptera (29 species), Coleoptera (15 species), Odonata (10 species), Hymenoptera (7 species), Hemiptera and Diptera (5 species each), and Mollusca, Oligochaeta and Homoptera (1 species each). Comparison of number of species of invertebrates in khajjiar lake area and known from Himachal Pradesh (Mehta, 2005) reveal the presence of 6.6% of the invertebrate fauna of the state is in Khajjiar area.

A total of 121 species of insects belonging to 108 genera spread over 28 families and 8 orders are present in Khajjiar area. It forms around 55% of the total fauna of the area and 98% of the invertebrates. Of these, Lepidoptera (49 species) is the most dominant insect order in the area, followed by Orthoptera (29 species), Coleoptera (15 species), Odonata (10), Hymenoptera (7), Hemiptera and Diptera (5 each), and Homoptera (1 species).

A total of 100 species of vertebrates belonging to 83 genera, spread over 49 families and 22 orders, representing around 45% of the total fauna are present in Khajjiar area. Classwise analyses of data reveals that Aves dominates the fauna with 77 species belonging to 62 genera followed by Mammalia (16 species under 14 genera), Reptilia (4 species belonging to 4 genera), Amphibia (2 species spread over 2 genera) and Pisces (single species under a single genus). It has been further observed that Khajjiar area supports a good population of some vertebrate species like Carp (*Cyprinus carpio*), Himalayan Toad (*Bufo himalayanus*), Kashmir Rock Lizard (*Laudakia tuberculata*), Cattle Egret (*Bubulcus ibis*), Black Kite (*Milvus migrans*), Himalayan Griffon(*Gyps himalayensis*), Blue Rock Pigeon (*Columba livia*), Oriental Turtle-Dove (*Streptopelia orientalis*), Slaty-headed Parakeet (*Psittacula himalayana*), Common Cuckoo (*Cuculus canorus*), White-breasted Kingfisher (*Halcyon smyrnensis*), Himalayan Pied Woodpecker (*Dendrocopos himalayensis*), Himalayan Bulbul (*Pycnonotus leucogenys*), Grey Bushchat (*Saxicola ferrea*), Streaked Laughingthrush (*Garrulax lineatus*), Rufous-bellied Niltava (*Niltava sundara*), Great Tit (*Parus major*), House Sparrow (*Passer domesticus*), Common Myna (*Acridotheres tristis*), Jungle Crow (*Corvus macrorhynchos*), Rhesus Monkey (*Macaca mulatta), Hanuman Langur (Semnopithecus ajax*),Himalayan Fox (*Vulpes vulpes*),Yellow throated Martin (*Martes flavigula*), Black Bear (*Ursus thibetanus*), Barking Deer (*Muntiacus muntjac*),Goral (*Nemarnhdus goral*), Flying Squirrel (*Petaurista petaurista*), House Mouse (*Mus musculus*) etc.

A single species of Pisces, i.e. *Cyprinus carpio,* Linnaeus is also presentin the Khajjiar lake. It appears to be introduced in the lake as also revealed by the locals. Two different varietiesC. *c.communis* Linnaeus (Common Carp) and *C. c.specularis* Lacepeds (Mirror Carp) are found there in the lake. **Only two species of Amphibia belonging to two different families are there in Khajjiar pasture.** But four species of reptiles belonging to four genera are there spread over thre efamilies and one order.

There is a presence of 77 species of birds belonging to 62 genera, 12 orders and 31 families. Birds represent 77% of the vertebrate and 34.5% of the total fauna of the Khajjiar lake. In this area, Muscicapidae is the most represented family with 22 species belonging to 15 genera followed by Accipitridae and Corvidae (6 species each), and Paridae, Phasianidae, Columbidae and Picidae (3 species each). Khajjiar lake and surrounding area supports 20 such species of birds which are resident and rest 57 are seasonal-local and long range migrants. The birds placed under resident category include critically endangered Indian White-backed Vulture and Red-headed Vulture. Of the 57 species, 35 are seasonal-local migrants, four are winter visitors and 10 are summer visitors. Moreover, Khajjiar Lake supports eight such species which showsthewinter and the summer influx. Of these, six show summer influx, whereas, thewinter influx is shown by only two species.

16 species of mammals belonging to 14 genera, 12 families and 6 orders are present in the Khajjiar wildlife sanctuary. Order wise analyses of the data reveals that order Carnivora supports maximum of 6 species followed by Artiodactyla (3 species), and Chiroptera, Primates and Rodentia (2 species each). Moreover, a single species belonging to order Insectivora is also present. Family wise analysis of data reveals that families Vespertilionidae, Cercopithecidae, Mustelidae and Bovidae are represented by two species each whereas, Ursidae, Hynaeidae, Felidae, Cervidae, Sciuridae, Muridae, Soricidae and Canidae are represented by one species each. Of the sixteen species of mammals, 11 species belong to the category of large mammals and 5 species to the category of small mammals.

Nine species are listed as threatened in Convention in Trade of Endangered Species (CITES) under different schedules. Five species namely *Semnopithecus ajax, Ursus thibetanus, Panthera pardus, Naemorhedus sumatraensis*and *Naemorhedus goral* are placed in schedule I, *Macaca mulatta* in Schedule II and *Vulpes vulpes, Martes flavigula* and *Mustela sibrica* under schedule III. Out of a total of sixteen species, thirteen are placed under Indian Wildlife Protection Act 1972. Two species *Panthera pardus* and *Naemorhedus sumatraensis*are kept under schedule I. Same species areconsidered as vulnerable species according to National Red Data. Mammals are present in all three parts of khajjiar area

The above account shows that the bio-diversity in Khajjiar lake area, like Himachal Pradesh is very rich and diversified, but, in recent years, Khajjiar in particular and the state in general have come under a strong threshold of development. Natural ecosystems/habitats have been over-exploited and even destroyed by the rapidly increasing human population and tourist inflow. A number of endemic and restricted range species found in the area/region are facing threat to their existence. Keeping in view the imminent dangers of extinction to bio-diversity some measures are proposed.

Keeping in mind the eutrophication of the lake, entry of stray and domesticated animals to the Khajjiar meadow should be immediately checked. Proper management of horse dung is immediately needed. Proper management of non-biodegradable as well biodegradable waste like plastic and bottles (plastic and glass) is required. There should be some check on the entry of tourists keeping in view the huge population of tourists visiting the meadow every year. Proper management of waste material from hotels is urgently required. Proper guidelines for tourism sports in Khajjiar meadow should be prepared. Keeping in view the disturbance created to the wildlife by vehicles in the form of horn, air pollution, noise pollution etc, traffic laws should be strictly implemented on all the vehicles entering the sanctuary area.

Literature Cited

Abdar, M.R., Patil, S.R. and Jadhav, B.V. (2006). The birds of lake Morana district Sangli (M.S.) in India. In: *Biodiversity and Environment* (eds.: Pandey B.N. and Kulkarni G.K.). A.P.H. Pub., New Delhi, 87-94.

Acharjee, M.N. and Kriplini, M.B. (1951). On a collection of reptilian and Batrachia from the Kangra and Kullu valleys, Western Himalayas. *Records of Indian Museum*, **44**, 175-184.

Agrawal, H.P. (1975 a). A check list of molluscs of Himachal Pradesh, India. Part-I. *Cheetal*, **16**, 38-41.

Agrawal, H.P. (1975 b). A check list of molluscs of Himachal Pradesh, India. Part-II. *Cheetal*, **16**, 59-60.

Agrawal, H.P. (1976). Aquatic and Amphibious molluscs of Himachal Pradesh, India. *Records of Zoological Survey of India*, **71**, 129-142.

Agrawal, H.P. (1977). *New records of land molluscs from Himachal Pradesh, India. Newsletter*, Zoological Survey of India, **3**, 345-346.

Agrawal, H.P. (1979). Some land molluscs from Himachal Pradesh. *Indian Journal of Zoology*, **18**, 131-135.

Agrawal, V.C. (1998). Mammalia. In: *Faunal Diversity in India*. Zoological Survey of India, Kolkata, 459-469.

Aitken, S. (2011). The UN Decade for Biodiversity 2011–2020. *Biodiversity*, **12**, 143.

Alexeev, A.V. (2009). New Jewel Beetles (Coleoptera: Buprestidae) from the Cretaceous of Russia, Kazakhstan, and Mongolia. *Paleontological Journal*, **43**, 277–281.

Alfred, J. R. B. and Ramakrishna. (2004). *Collection, Preservation and Identification of Animals*. Zoological Survey of India, Kolkata, p 310.

Alfred, J.R.B., Das, A.K. and Sanyal, A.K. (1998). *Faunal diversity in India*. Zoological Survey of India , Kolkata, p 495.

Ali, S. 1949. *Indian Hill Birds*. Oxford University Press, Bombay, p 188.

Ali, S. and Ripley, S.D. (1983 b). *A Pictorial Guide to the Birds of the Indian Subcontinent*. Bombay Natural History Society/Oxford University Press, New Delhi, p 177.

Allen, G.M. (1908). Notes on Chiroptera. *Bulletin of the Museum of Comparative Zoology*, **52**, 25–61.

Allport, G., Ausden, M., Hayman, P.V., Robertson, P.A. and Wood, P.N. (1988). The birds of Gola forest, Sierra Leone. *International Center for Birds of Prey Study Report* 38, Cambridge.

Alonso, A., Dallmeier, F., Granek, E. and Raven, P. (2001). *Biodiversity: Connecting with the Tapestry of life*. Smithsonian Institute. Washington D. C., U.S.A, p 31.

Ananthakrishnan, T.N. and Shivaramakrishnan, K.G. (2006). *Animal biodiversity: patterns and processes*. Scientific Publishers, Jodhpur, p 264.

Anderson, J.C. (1889). Sporting rambles round about Shimla. *Journal of Bombay Natural History Society*, **4**, 56-66.

Andheria, A. P. (2001). Diversity of butterflies near a pool in the Sanjay Gandhi National Park, Mumbai. *Journal of Natural History*, **98**, 302-303.

Anjos, L.D., Collins, C.D., Holt, R.D., Volpato, G.H., Mendonça, L.B., Lopes, E.V., Bocon, B., Bisheimer, M.V., Serafini, P.P. and Carvalho, J. (2011). Bird species abundance–occupancy patterns and sensitivity to forest fragmentation: Implications for conservation in the Brazilian Atlantic forest. *Biological Conservation*, **144**, 2213–2222.

Annandale, N. (1907). The distribution of *Bufo andersonii*. *Records of Indian Museum*, **1**, 171-172.

Arora, G.S. (1990). *Collection and preservation of animals: Lepidoptera*. Zoological Survey India, Kolkata, 131-138.

Arora, G.S. (1994). Lepidoptera: Butterflies. In: *Fauna of Rajaji National Park*. Zoological Survey of India, Kolkata, 245-300.

Arora, G.S. (1995). Lepidoptera: Rhopidorae. In: *Fauna of Western Himalaya, Uttar Pradesh*. Zoological Survey of India, Kolkata, 61-73.

Arora, G.S. (2000). Lepidoptera: Butterflies. In: *Fauna of Renuka Wetland, Wetland ecosystem Series-2*. Zoological Survey of India, Kolkata, 105-120.

Arora, G.S., Mehta, H.S. and Walia, V.K. (2005). Insecta: Lepidoptera (Butterflies). In: *Fauna of Western Himalaya (Part 2)*. Zoological Survey of India, Kolkata, 157-180.

Arrow, G.J. (1910). *The Fauna of British India including Ceylon and Burma*. Coleoptera: Lamellicornia, Scarabaeidae, I. Cetoninae and Dynastinae. Taylor and Francis, London, p 322.

Arrow, G.J. (1931). *The Fauna of British India including Ceylon and Burma*. Coleoptera: Lamellicornia, Scarabaeidae, III. Coprinae. Taylor and Francis, London, p 428.

Bahuguna, A. (2008). Altitudinal Variations in Morphological Characters of *Laudakia tuberculata* Hardwicke and Gray, 1827 from Western Himalayas (Uttarakhand), India. *Russian Journal of Herpetology*, **15**, 207-211.

Baqri, Q.H. (1998). Nematoda. In: *Faunal Diversity in India*. Zoological Survey of India, Kolkata, 85-92.

Barman, R.P. (1998). Pisces. In: *Faunal Diversity in India*. Zoological Survey of India, Kolkata, 418-426.

Basu, C.R. (1986). Chrysomelidae (Coleoptera) of Silent Valley, Kerala. *Records of Zoological survey of India*, Koltata, **84**, 143-155.

Basu, C.R. and Halder, S.K. (1987). Insecta: Coleoptera, Chrysomelidae. *Fauna of Orissa (Part-1)*, Zoological Survey of India Publication, Kolkata, 213-240.

Belsare, D.K. (2007). *Introduction to Biodiversity*. APH Publication Corporation, New Delhi, p 264.

Besten, J.W. (2004). *Birds of Kangra*. Moonpeak Publishers, Dharamsala and Mosaic Books, New Delhi, p 173.

Betrem, J. G. (1928). Monographic der Indo-Australischen Scoliiden mit zoogeographischen Betrachungen. *Treubia*, **9**, 338.

Bhalla, O.P. and Pawar, A.D. (1977). *A survey of insect and non-insect pests of economic importance in Himachal Pradesh*. In Monograph, Department of Entomology-Zoology, College of Agriculture, Chambaghat, Solan, 1-80.

Bhardwaj, A.K. and Thakur, J.R. (1973). Record of snail *Bensonia monticola* Hutton, as pest of beans. *Current Science*, **42**, 769.

Bhardwaj, S. (2003). Status assessment of Birds of urban Shimla. *Ph.D. thesis, Himachal Pradesh University*, Shimla, India.

Bhargav, K.V. and Uniyal, V.P. (2008). Communal Roosting of Tiger Beetles (Cicinelidae: Coleoptera) in the Shivalik Hills, Himachal Pradesh, India. *Cicindela*, **40**, 1-12.

Bhasin, G.D., Roonwal, M.L. *et al.*, (1953). Odonata. In: A systematic catalogue of the main identified entomological collections at the Forest Research Institute, Dehradun, 9-21, 65-79.

Bhat, H.R., Kulkarni, S.M. and Mishra, A.C. (1983). Records of Mesostigmata, Ereynetidae and Pterygosomidae (Acarina) in Western Himalayas, Sikkim and hill districts of West Bengal. *Journal of the Bombay Natural History Society*, **80**, 91–110.

Bhatanagar, G.K. (1973). On a collection of fish from Bhakra reservoir, Sutlej river and closely associated waters. *Journal of Inland Fisheries Society of India*, **5**, 134-136.

Bhatnagar, C., Koli, V. K. and Sharma, S. K. (2010). Summer Diet of Indian Giant Flying Squirrel Petaurista Philippensis (Elliot) in Sitamata Wildlife Sanctuary, Rajasthan, India. *Journal of the Bombay Natural History Society*, **107**, 183-188.

Bhowmick, H.K. (1985 a). Outline of distribution with an index catalogue of Indian grasshoppers (Orthoptera: Acridoidea). *Records of Zoological Survey of India*, **78**, 1-51.

Bhowmick, H.K. (1985 b). Contribution to the Gryllidae fauna of the Western Himalayas (Orthoptera: Gryllidae). *Records of Zoological Survey of India*, **73**, 1-74.

Bhowmick, H.K. (1986). (Orthoptera: Acrididae). In: *Grasshopper fauna of West Bengal, India*. Zoological Survey of India, Kolkata, **14**, 1-180.

Bhowmick, H.K. and Halder, P. (1983). Preliminary distributional records with remarks on little known species of Acrididae (Orthoptera: Insecta) from the western Himalayas (Himachal Pradesh). *Records of Zoological Survey of India*, **81**, 167-191.

Bhowmick, H.K. and Halder, P. (1984). Preliminary distribution with remarks on little known species of Acrididae (Orthoptera: Insecta) from the western Himalayas, (Himachal Pradesh). *Records of Zoological Survey India,* **18**, 167-191.

Bhowmick, H.K. and Rui, K.N. (1982). Notes on a collection on grasshopper (Orthoptera: Acrididae) from Shiwalik hills. *Indian Museum Bulletin*, **17**, 48-54.

Biggins, R.G., Neves, R.J. and Dohner, C.K. (1995). *Draft national strategy for the conservation of native freshwater mussels.* U.S. Fish and Wildlife Service, Washington, DC, p 26.

Bingham, C.T. (1897). *The Fauna of British India including Ceylon and Burma.* Hymenoptera-1. Taylor and Francis, London, p 579.

Bingham, C.T. (1903). *The Fauna of British India including Ceylon and Burma.* Hymenoptera-II. Taylor and Francis, London, p 506

Bingham, C.T. (1905). *The Fauna of British Indian including Ceylon and Burma.* Butterflies-1. Taylor and Francis Ltd. London, p 537.

Bingham, C.T. (1907). *The Fauna of British India including Ceylon and Burma.* Butterflies-2. Taylor and Francis, Ltd. London, p 480.

Bird Life International (2001). Threatened Birds of Asia: The Birdlife International Red Data Book (Vol. 1& 2) (Series eds.: Collar, N.J., Andreev, A.V., Chan, S., Crosby, M.J., Subramanya, S. and Tobias, J.A.) *Birdlife International,* Cambridge, UK, p 3038.

Bisht, R., Pandey, H., Bharti, D. and Kaushal, B. R. (2003). Population dynamics of earthworms (Oligochaeta) in cultivated soil of central Himalayan Tarai region. *Tropical Ecology*, **44**, 221-226.

Blanford, W. T. and Godwin-Austen, H. H. (1908). *The Fauna of British India Including Ceylon and Burma.* Mollusca- I. Taylor and Francis, London, p 331.

Bolivar, I. (1902). Les Orthopteres de St. Joseph's College, a Trichinolopy. *Annales de la Societe Entomologique de France*, 70, 580-635.

Bolivar, I. (1909). Orthoptera (Fam. Acrididae. Subfam. Pyrgomorphinae) Genera Insectorum, Brussels. *Faasc*, **90**, p 58.

Bolivar, I. (1914). Estudios Entomologicos. II. Los Truxalinos del antiguo Mundo. *Trab. Mus. Nac. Cienc. Nat. Madrid*, **16**, 278-412.

Borror, K.J., Delong, D. M. and Triplehonrn, C.A. (1981). *An introduction to the study of insects.* Saunders College Publishing, Holt, Rinehart and Winston, 281-285.

Bouchet, P. (1996). Extinction and preservation of species in the tropical world: What future for Mollusks? *American Conchologist*, **20**, 20-24.

Boulenger, G.A. (1920). A monograph of south Asian Pipuan, Melanesian and Australian frog of genus *Rana*. *Records of Indian Museum*, **20**, 1- 226.

Brown, B.V., Borkent, A., Cumming, J.M., Wood, D.M., Woodley, N.E., and Zumbado, M. (Editors) (2009). *Manual of Central American Diptera.* Volume 1 NRC Research Press, Ottawa.

Bruggen, A.C.V. (1995). Biodiversity of Mollusca: Time for a new approach . In: *Biodiversity and Conservation of the Mollusca.* eds.: van Bruggen A. C., S. M. Wells, T. C. M. Kemperman, Oegstgeest-Leiden, Backhuys, the Netherlands, 1–19.

Brunetti, E. (1917). Diptera of the Simla District. *Records of Indian Museum*, **13**, 59-101.

Burnham, K.P., Anderson, D.R. and Laake, J.L. (1980). *Estimation of density from line transect sampling of biological populations*. Wildlife Monograph No. 72, The Wildlife Society, USA, p 202.

Cantile, K. (1962). *The Lycaenidae portion (except Arthopola group) of Brigadier Evans the identification of Indian butterflies*, 1932 (India, Pakistan, Ceylon, burma), p 159.

Chakraborty, S., Mehta, H.S. and Pratihar, S. (2005). Mammals. In: *Fauna of Western Himalaya (Part 2)*. Zoological Survey of India, Kolkata, 341-359.

Chanda, (2002). *Hand book. Indian Amphibians*. Zoological Survey of India, Kolkata, p 335.

Chanda, S.K. (1998). Amphibia. In: *Faunal Diversity in India*. Zoological. Survey of India, Kolkata, 427- 433.

Chandra, K. (2005). Insecta: Coleoptera: Scarabaeidae. In: *Fauna of Western Himalaya (Part 2)*. Zoological Survey of India, Kolkata, 141-155.

Chandra, K. and Uniyal, V. P. (2007). On a collection of Pleurostict Scarabaeidae (Coleoptera) from Great Himalayan National Park, Himachal Pradesh. *Zoos' Print Journal*, **22**, 2821-2823.

Chandra, M. (1983). Additions to the Odonata fauna of districy Solan, Himachal Pradesh. *Fraseria*, **5**, 17-18.

Chaudhuri, P. S., Nath, S. and Paliwal, R. (2008). Earthworm population of rubber plantations (Hevea brasiliensis) in Tripura, India. *Tropical Ecology*, **49**, 225-234.

Chauhan, R. (1998). *Himachal Pradesh-A perspective*. Minerva Book House, Shimla.

Chavan, V., Watve, A.V., Londhe, M.S., Rane, N.S., Pandit, A.T. and Krishanan, S. (2004). Cataloguing Indian biota: the electronic catalogue of known Indian fauna. *Current Science*, **87**, 749-763.

Chavhan, A. and Pawar, S.S. (2011). Distribution and Diversity of Ant Species (Hymenoptera: Formicidae) in and Around Amravati City of Maharashtra, India. *World Journal of Zoology*, **6**, 395-400.

Chopard, L. (1969). *The Fauna of India and adjacent countries on Orthoptera (Grylloidea)*. Zoological Survey of India, Kolkata, **2**, 1-415

Crosskey, R.W. (1965). The identification of African Simuliidae (Diptera) living in phoresis with nymphal Ephemeroptera, with special reference to Simuiium berneri Freeman. *Proceedigs of the Royal entomological Society of London*, **40**, 118-24.

Daniel, J.C. (1989). Bird sanctuary at Kehim. *Newsletter for Birdwatchers*, **29**, 9.

Dario, F.R. and Vincenzo, M.C.V.D. (2011). Avian diversity and relative abundance in a restinga forest of Sao Paulo, Brazil. *Tropical Ecology*, **52**, 25-33.

Das, Abhijit, Saikia, U., Murthy, B.H.C.K., Dey, S. and Dutta, S.K. (2009). A herpetofaunal inventory of Barail Wildlife Sanctuary and adjacent regions, Assam, northeastern India. *Hamadryad*, **34**, 117-134.

Das, D. and Saikia, P.K. (2007). Present status of soft shell turtles (family Trinonychidae) in Barpete and its surroundings areas, Barpete, Assam. *Reptile Rap*, **8**, 3-4.

Das, I. (1994). The reptiles of South Asia, checklist and distribution summary. *Hamadryad*, **19**, 15-40.

Datta, M. (1998). Diptera. In: *Faunal Diversity in India*. Zoological Survey of India, Kolkata, 284-293.

Datta, M. and Chakraborti, M. (1983). On a collection of flower flies (Diptera: Syrphidae) with new records from Jammu and Kashmir, India. *Records of Zoological Survey of India*, **82**, 231–52.

Datta, M. (1998). Diptera. In: *Faunal Diversity in India*. Zoological Survey of India, Kolkata, **91**, 49-51

Davis, G.M., Subba Rao, N.V. and Hoagland, K.E. (1986). In search of *Tricula*. *Tricula* defined and a new genus described. *Proceedings of the Academy of Natural Sciences, Philadelphia*, **138**, 426-442.

Day, F. (1889 a). *The Fauna of British Indian including Ceylon and Burma*. Fishes-I. Taylor and Francis Ltd. London, p 548.

Day, F. (1889 b). *The Fauna of British Indian including Ceylon and Burma*.Fishes-II. Taylor and Francis Ltd. London, p 449.

De Niceville, L. (1886). *The butterflies of India, Burma and Ceylon*. Taylor and Francis Ltd. London, **2**, 332.

Devroy, M.K. (2008). Crustacea. In: *Fauna of Pin Valley National Park*. Zoological Survey of India publication, Kolkata, **34**, 23-24.

Dhanze, R. and Dhanze, J.R. (2004). *Fish diversity of Himachal Pradesh. In: Fish diversity in protected habitats*. NATCON, Publication, U.P.

Dhindsa, M.S., Sandhu, P.S. and Sandhu, J.S. (1991). Some additions to the checklist of birds of Punjab. *Pavo*, **28**, 23-28.

Dinesh, K.P., Radhakrishnan, C., Gururaja, K.V. and Bhatta, G.K. (2009). An annotated checklist of Amphibia of India with some insights into the patterns of species discoveries, distribution and endemism. *Records of Zoological Survey of India*, 302, 1-153.

Dirsh V. M. (1956). Preliminary revision of the genus Catantops Schaum and review of the group Catantopini (Orthoptera : Acrididae). *Publ. cult. Cia. Diamante Angola, Lisbon*, **28**, 11-150.

Distant,W.L. (1904). *The Fauna of British India including Ceylon and Burma*. Rhynchota 2. Taylor and Francis Ltd. London, p 487.

Distant,W.L. (1906). *The Fauna of British India including Ceylon and Burma*. Rhynchota 3. Taylor and Francis Ltd. London, p 491.

Distant,W.L. (1908). *The Fauna of British India including Ceylon and Burma*. Rhynchota 4. Taylor and Francis Ltd. London, p 501.

Distant,W.L. (1911). *The Fauna of British India including Ceylon and Burma*. Rhynchota 5. Taylor and Francis Ltd. London, p 362.

Distant,W.L. (1916). *The Fauna of British India including Ceylon and Burma*. Rhynchota 6. Taylor and Francis Ltd. London, p 248.

Distant,W.L. (1918). *The Fauna of British India including Ceylon and Burma.* Rhynchota 7. Taylor and Francis Ltd. London, p 210.

Distant,W.L. (1902). *The Fauna of British India including Ceylon and Burma.* Rhynchota 1. Taylor and Francis Ltd. London, p 424.

Dobson, G.E. (1873). Description of a new species of *Vespertilio* from northwestern Himalaya. *Journal of the Asiatic Society of Bengal,* **42**, 205–206.

Dodsworth, P.T.L. (1910 a). Notes relating to the distribution, habits and nidification of *Certhia himalayana* Vigors (The Himalayan Tree-creeper) in and around Simla and the adjacent ranges. *Journal of the Bombay Natural History Society,* **20**, 463-467.

Dodsworth, P.T.L. (1910 b). The Himalayan Greenfinch. *Journal of the Bombay Natural History Society,* **20**, 517.

Dodsworth, P.T.L. (1911 a). Protection of wild birds in India and traffic in plumage. *Journal of the Bombay Natural History Society,* **20**, 1103-1114.

Dodsworth, P.T.L. (1911 b). Crow and its food. *Journal of the Bombay Natural History Society,* **21**, 248-249.

Dodsworth, P.T.L. (1911 c). Occurrence of *Hemilophus pulverulentus* Temm, the Great Slaty Woodpecker in the neighbourhood of Simla, N.W. Himalayas. *Journal of the Bombay Natural History Society,* **21**, 263.

Dodsworth, P.T.L. (1912 a). The Crag Martin (*Ptyonoprogne rupestris*). *Journal of the Bombay Natural History Society,* **21**, 660-661.

Dodsworth, P.T.L. (1912 b). Extension of the habitat of the Common Kingfisher (*Alcedo ispida*). *Journal of the Bombay Natural History Society,* **21**, 661.

Dodsworth, P.T.L. (1912 c). Extension of the habitat of the Brahminy Kite (*Haliastur indus*). *Journal of the Bombay Natural History Society,* **21**, 665-666.

Dodsworth, P.T.L. (1912 d). Distribution, habits, and nesting of the Himalayan Greenfinch, *Hypacanthis spinoides* Vigors. *Journal of the Bombay Natural History Society,* **21**, 1075-1080.

Dodsworth, P.T.L. (1912 e). Occurrence of the Common Peafowl, *Pavo cristatus* Linnaeus in the neighbourhood of Simla, N.W. Himalayas *Journal of the Bombay Natural History Society,* **21**, 1082-1083.

Dodsworth, P.T.L. (1912 f). Question whether *Gyps fulvus* Gmelin, the Griffon, occurs in the Himalayan districts of the Punjab. *Journal of the Bombay Natural History Society,* **21**, 1331-1332.

Dodsworth, P.T.L. (1912 g). A Kite's larder. *Journal of the Bombay Natural History Society,* **21**, 1332-1333.

Dodsworth, P.T.L. (1913 a). Notes on the Vultures found in the neighbourhood of Simla and adjacent ranges of the Himalayas. *The International Journal of Avian Science (Ibis),* **1**, 534-544.

Dodsworth, P.T.L. (1913 b). Occurrence of the Red-tailed Chat, *Saxicola chrysopygia* De Filippi in the vicinity of Simla. *Journal of the Bombay Natural History Society,* **22**, 196.

Dodsworth, P.T.L. (1913 c). Notes on some mammals found in Simla districts,the Simla hill states, and Kalka and adjacent country. *Journal of the Bombay Natural History Society*, **22**, 726–748.

Dodsworth, P.T.L. (1914). Occurrence of the White-browed Bush-Robin, *Ianthia indica* (Vieill.) in the North-West Himalayas. *Journal of Bombay Natural History Society*, **22**, 795-796.

Doherty, W. (1886). A list of butterflies taken in Kumaon. *Journal of Asiatic Society of Bengal*, **55**, 103-140.

Dubois, A. (1975). Un nouvcau sous-genre (*Paa*) et trios nouvelles especes du genera *Rana*. Remarques sur la phylogenie des Ranides (Amphibiens, Anoures). *Bailey Museum of Natural History*, **231**, 1093-1145.

Dutta, S. K. (1997). *Amphibians of India and Sri Lanka (checklist and bibliography)*. Odyssey Publishing House, Bhubaneswar, India, p 342.

Emberton, K. C. (1995 a). Land-snail community morphologies of the highest diversity sites of Madagascar, North America and New Zealand, with recommended alternatives to height-diameter plots. *Malacalogia*, **36**, 43-46.

Emberton, K. C. (1995 b). On the endangered biodiversity of Madagascan land snails. In: *Biodiversity and Conservation of the mollusca* (Eds. A. C. V Bruggen, S. Wells and T. C. M. Kemperman), 69-89.

Emberton, K. C. (1996). Conservation priorities for forest-floor invertebrates of the southeastern half of Madagascar: evidence from two land snails. *Biodiversity and Conservation*, **5**, 729-741.

Evans, W.H. (1949). *A Catalogue of the Hesperiidae from Europe, Asia and Australia in the British Museum*. Trustees of B.M. London, p 502.

Fowler, W.W. (1912). *The Fauna of British India including Ceylon and Burma*. Coleoptera: General Introduction, Cicindelidae and Paussidae. Taylor and Francis Ltd. London, p 529.

Fraser, F.C. (1933). *Fauna of British India including Ceylon and Burma*. Odonata- I. Taylor and Francis Ltd. London, p 423.

Fraser, F.C. (1934). *Fauna of British India including Ceylon and Burma*. Odonata- II. Taylor and Francis Ltd. London, p 338.

Fraser, F.C. (1936). *Fauna of British India including Ceylon and Burma*. Odonata- III. Taylor and Francis Ltd. London, p 461.

Frome, N.F. (1946). Birds noted in the Mahasu-Narkanda-Baghi area of the Shimla Hills. *Journal of the Bombay Natural History Society*, **46**, 308-316.

Gahan, C.J. (1906). *The Fauna of British India including Ceylon and Burma*. Coleoptera: Cerambycidae- 1. Taylor and Francis Ltd. London, p 329.

Ganesh, T., Priyadarsanan, D.R., Devy, M.S., Aravind, N.A. and Rao, D. (2002). Assessment of biodiversity of lesser-known and functionally important groups in Rajiv Gandhi Nagarahole National park. Report: Karnataka Forest Department, Bangalore, India.

Ganguli, U. (1967). Birds of Simla in autumn. *Newsletter for Birdwatchers*, **7**, 4-6.

Garson, P.J. (1983). The Cheer Pheasant (*Catreus wallichii*) in Himachal Pradesh, Western Himalayas: An update. *Journal of World Pheasant Association*, **8**, 29-39.

Gaston, A.J. and Pandey, S. (1987). Sighting of Rednecked Grebes (*Podiceps grisegena*) on the Pong Dam Lake, Himachal Pradesh. *Journal of Bombay Natural History Society*, **84**, 676-677.

Gaston, A.J. and Singh, J. (1980). The status of the Cheer Pheasant (*Catreus wallichii*) in the Chail Wildlife Sanctuary, Himachal Pradesh. *Journal of World Pheasant Association*, **5**, 68-73.

Gaston, A.J., Garson, P.J. and Hunter, M.L. Jr. (1981). Present distribution and status of Pheasants in Himachal Pradesh, Western Himalayas. *World Pheasent Association Journal*, **6**, 10-30.

Gaston, A.J., Garson, P.J. and Pandey, S. (1993). Birds recorded in the Great Himalayan National Park, Himachal Pradesh, India. *Forktail*, **9**, 45-57.

Gates G.E. (1972). Burmese earthworms. An introduction to the systematics and biology of megadrile oligochaetes with special reference to Southeast Asia. *Trans: American Philosophical Society*, **62**, 10326.

Gay, T., Kehimkar, I.D. and Punetha, J.C. (1992). *Common butterflies of India*. Published for world wild fund for Nature-India, Oxford University Press, New Delhi, p 67.

Ghosh, D., Debnath, N. and Chakrabarti, S. (1991). Predators and parasites of Aphids from North West and Western Himalayas III. Twenty five species of Coccinellidae (Coleoptera: Insecta) from Garhwal and Kumaon ranges. *Records of Zoological Survey India, Kolkata*, **88**, 177-188.

Ghosh, L.K., Biswas, B. and Ghosg, M. (2005). Insecta: Hemiptera. In: *Fauna of Western Himalaya (Part 2)*. Zoological Survey of India, Kolkata, 111-139.

Ghosh, R.K. (1998). Platyhelminthes. In: *Faunal diversity of India*, Zoological Survey of India, Kolkata, Kolkata, 50-55.

Ghosh, S.K. (1991). *Lesser know animal resources of India (Butterflies)*. Zoological Survey of India, Kolkata, 93-110.

Ghosh, S.K. and Chaudhary, M. (1997 a). Insecta: Lepidoptera: Hesperiidae. In: *Fauna of West Bengal*. Zoological Survey of India, Kolkata, 275-318.

Ghosh, S.K. and Chaudhary, M. (1997 b). Insecta: Lepidoptera: Pieridae. In: *Fauna of West Bengal*. Zoological Survey of India, Kolkata, 705-728.

Ghosh, S.K. and Chaudhary, M. (1998 a). Insecta: Lepidoptera: Hesperiidae. In: *Fauna of Meghalaya*. Zoological Survey of India. Kolkata, 269-341.

Ghosh, S.K. and Chaudhary, M. (1998 b). Insecta: Lepidoptera: Pieridae. In: *Fauna of Meghalaya*. Zoological Survey of India, Kolkata, 245-267.

Gillot, C. (2005). *Entomology*. Third addition, Springer Publication, The Netherland, 243-331.

Giri, S., Aryal, A., Koirala, R.K., Adhikari, B. and Raubenheimer, D. (2011). Feeding Ecology and Distribution of Himalayan Serow *(Capricornis thar)* in Annapurna Conservation Area, Nepal. *World Journal of Zoology*, **6**, 80-85.

Godwin-Austen, H.H. (1899). Address of the President. Appendix A. List of shells from Kashmir territory, South of the Pir Panjal and Kajnag ranges including the Murree Hills and Hazara. *Proceedigns of the Malacological Society of London*, **3**, 259-262.

Gomathi, N. and Singh, R. (2007). A note on the feeding habit of Indian soft shell turtle *Aspideretes gangeticus* in Keoladeo national park. *Reptile Rap*, **8**, 5-6.

Gong, J. and Harris, R. B. (2006). *The status of bears in China. Understanding Asian bears to secure their future*, Japan Bear Network, Ibaraki, Japan, 50-56.

Gopi Sundar, K.S. (2000). Distribution, demography and conservation status of the Indian Sarus Crane (*Grus antigone antigone*) in India. *Journal of the Bombay Natural History Society*, **97**, 319-339.

Goyal, A.K. and Arora, S. (Eds.) (2009). *Indian fourth National Report to the Convention on Biological diversity*. Ministry of Environment and Forests (Gol), New Delhi, 1-156.

Grimmett, R., Inskipp, C. and Inskipp, T. (1999). *Pocket Guide to the Birds of the Indian Subcontinent*. Oxford University Press, New Delhi, p 384.

Gruber, U. (1981). Notes on the Herpetofauna of Kashmir and Ladakh. *British Journal of. Herpetology*, **6**, 145-150.

Gude, G. K. (1914). *The Fauna of British India- Including Ceylon and Burma*. Mollusca II. Taylor and Francis Ltd. London., 520.

Gude, G. K. (1921). *The Fauna of British India- Including Ceylon and Burma*. Mollusca III. Taylor and Francis Ltd. London, 386.

Gupta, I.J. (1997a). Lepidoptera: Nymphalidae. In: *Fauna of Delhi*. Zoological Survey of India, Kolkata, 409-414.

Gupta, I.J. (1997b). Insecta: Lepidoptera: Lycaenidae. In: *Fauna of West Bengal*. Zoological Survey of India, Kolkata, 429-489.

Gupta, I.J. (1997c). Insect: Lepidoptera: Amathusiidae: Acracidae: Nymphalidae: Riodinidae. In: *Fauna of West-Bengal*. Zoological Survey of India, Kolkata, 533-612.

Gupta, I.J. and Shukla, J.P.N. (1987). Butterflies from Bastar district (M.P.) *Records of Zoological Survey of India*, **106**, 1-74.

Gupta, I.J. and Shukla, J.P.N. (1988). Studies on the butterflies of Arunachal Pradesh and adjoining areas, India (Lepidoptera: Acraeidae, Satyaridae, Nymphalidae, Riodinidae, Lycaenidae). *Records of Zoological Survey of India*, **109**, 1-115.

Gupta, I.J. and Thakur, R.K. (1990). Lepidoptera fauna of Gujarat, India. *Records of Zoological Survey of India*, 229-24.

Gupta, S. K. D., Mazumdar, A and Chaudhuri, P. K. (2008). Biting Midges of the Genus *Palpomyia* Meigen (Diptera: Ceratopogonidae) in India. *Bonner Zoologische Beitrage*, **56**, 43–48.

Gupta, S.K. (1994). Hymenoptera (Insecta) In: *Fauna of Rajaji National Park*. Zoological Survey of India, Kolkata, 301-307.

Gupta, S.K. (1995). Hymenoptera. In: *Fauna of Western Himalayas, Uttar Pradesh*. Zoological Survey of India, Kolkata, 81-89.

Gupta, S.K. (1997). Hymenoptera. In: *Fauna of Nanda Devi Biosphere Reserve*. Zoological Survey of India, Kolkata, 97-104.

Gupta, S.K. (2004). Insecta: Hymenoptera: Aculeata. In: *some selected fauna of Gobind Pashu Vihar*, Zoological Survey of India, Kolkata, **18**, 21-28.

Gupta, S.K. (2008). Insecta: Hymenoptera: Aculeata. (Vespidae and Sphecidae). In: *Fauna of Pin Valley National Park*, Zoological Survey of India publication, Kolkata, **34**, 61-64.

Guptha, M.B., Rao, P.V.C., Ramalingam, G., Prasad, P.N., Kishore, S. And Rajasekher, M. (2011). The water birds of Nelapattu bird sanctuary Andra Pradesh, India. *World Journal of Zoology*, **6**, 249-254.

Hamlyn, 1969. Lepidoptera. In: *Faunal diversity in India*. Zoological Survey of India, Kolkata, 312-318.

Hampson, G. F. (1918). *Catalogue of the Lepidoptera Phalaenae in the collection of the British Museum, Supplement* Vol.1, Amatidae, Nolinae & Lithosiinae. Trustees of the British Museum of Natural History, London, 824.

Haribal, M. (2001). Over wintering population of *Danaus* (*Salathura*) *genutia* in tiger valley of Sanjay Gandhi National Park, Mumbai, Maharashtra. *Journal of the Bombay Natural History Society*, **98**, 469-472.

Hemming, F. (1967). The generic names of the butterflies and their type species (Lepidoptera: Rhopalocera). *Bulletin of the British Museum of Natural History, Entomology*, **9**, 509.

Henry, G.M. (1940). New *and little known South Indian Orthoptera*. Royal Entomological Society of London, **90**, 497-540.

Holmes, P.R. (1986). The avifauna of the Suru River Valley, Ladakh. *Forktail*, **2**, 21-41.

Hope, F. W. (1837). *The coleopterists' manual, containing the lamellicorn insects of Linnaeus and Fabricius*. H. G. London, United Kingdom, p 121.

Hora, S.L. (1927). On a peculiar fishing implement from Kangra Valley, Punjab. *Journal and Proceedings of Asiatic Society of Bengal*, **22**, 81-84.

Hora, S.L. (1928). Hibernation and aestivation in gastropod molluscs. On the habit of a slug from Dalhousie (Westren Himalayas) with remarks on certain other species of Gastropod Molluscs. *Records of Indian Museum*, **30**, 357-373.

Hora, S.L., Mulik, G.M. and Khajuria, H. (1955). Some interesting features of the aquatic fauna of the Kashmir valley. *Journal of the Bombay Natural History Society*, **53**, 140-143.

Hwang, M.H. and Garshelis, D.L. (2007). Activity patterns of Asiatic black bears (Ursus thibetanus) in the Central Mountains of Taiwan. *Journal of Zoology* (London), **271**, 203-209.

Hwang, M.H. Garshelis, D.L. and Wang, Y. (2002). Diets of asiatic black bears in Taiwan, with methodological and geographical comparisons. *Ursus*, **13**, 111-125.

Ishwar, N.M., Chellam, R. and Kumar, A. (2001). Distribution of forest floor reptiles in the rainforest of Kalakad- Mundanthurai Tiger Reserve, South India., *Current Science*, **80**, 413-418.

IUCN (2008). IUCN Red List of Threatened Species. IUCN Species Survival Commission. Gland, Switzerland (www.iucnredlist.org).

Jacoby, M. (1908). *The Fauna of British India including Ceylon and Burma.* Coleoptera: Chrysomelidae, (Eupodes, Camptosomes, Cyclica). Taylor and Francis, London, **1**, 534 - 172.

Jairajpuri, M.S. (1993). Faunal diversity of Himalaya: Need for inventorization and constrains. In: *Himalayan biodiversity-conservation strategies*, GB Pant Institute of Himalayan Environment and Development, Almora, 57-64.

James, M. T. (1977). Family Calliphoridae. In Delfinado, M. D. and D. E. Hardy (eds.), A Catalog of Diptera of the Oriental Region, **3**, 526-556.

James, M.T. (1970). Family Calliphoridae, in Museu de Zoologia, Universidade de Sao Paulo, A Catalogue of the Diptera of the Americas south of the United States, **102**, 1-28.

Javed, S. (1996). Study on bird community structure of terai forest in Dudwa National Park. *Ph.D. thesis, Aligarh Muslim University,* Aligarh, India, p 149.

Jayaram, K. C. (1999). *The fresh water fishes of the Indian region.* Narendra Publishing House, Delhi, p 551.

Jayaram, K.C. (2006). *The Catfishes of India.* Narendra Publishing House, New Delhi, 383.

Jerdan,T.C. (1984) *A handbook of the mammals of India.* Mittal Publication Delhi, p 335.

Jhunjhunwala, S., Rahmani, A.R., Ishtiaq, F. and Islam, Z. (2001). The Important Bird Areas Programme in India. *Buceros,* **6**, 1-50.

Johal, M.S., tandon, K.K., Tyor, A.K. and Rawal, Y.K. (2002). Fish diversity in different habitats in the streams of lower western Himalayas. *Polish Journal of Ecology,* **50**, 45-56.

Jonathan, J.K. (1990). *Collection and preservation of animals: Hymenoptera.* Zoological Survey of India Publication, Kolkata, 174-150.

Jonathan, J.K. (1998). Hymenoptera. In: *Faunal Diversity in India.* Zoological Survey of India Publication, Kolkata, 326- 334.

Jonathan, J.K. (2005). Insecta: Hymenoptera: Ichneumonidae: *Fauna of Western Himalaya (Part-2).* Zoological Survey of India Publication, Kolkata, 191-220.

Julka J.M. (1981). Anthropochorous earthworms of Lahaul Valley (Himachal Pradesh with notes on their ecology. In: *Progress in Soil Biology and Ecology in India,* UAS Tech. Ser., University of Agricultural Sciences, Bangalore, **37**, 69-76.

Julka, J.M. (1988). *The Fauna of India and the adjacent countries, Megadrile Oligochaeta, Octochaetidae.* Zoological Survey of India, Kolkata, p 400.

Julka, J.M. (1998). Annelida. In: *Faunal Diversity in India.* Zoological Survey of India, Kolkata, 123-131.

Julka, J.M. and Mehta, H.S. (2000). *Fauna of Renuka wetland.* Zoological Survey of India, Kolkata, p 187.

Julka, J.M. and Paliwal, R. (1995). First record of *Microscolex phosphoreus* and *Malabaria levis* (Oligochaeta: Acanthodrillidae and Ocnerodrillidae) from India. *Megadrilogica,* **6**, 60-62.

Julka, J.M. and Paliwal, R. (2005). *Annelida: Oligochaeta. In: Fauna of Western Himalaya (Part 2)*. Zoological Survey of India, Kolkata, 53-60.

Kalsi, R.S. and Sodhi, N.S. (1986). Some birds of Chandigarh and surrounding areas. *Indian Journal of Poultry Science*, **21**, 162-163.

Karmegam, N. and Daniel, T. (2007). Effect of physicochemical parameters on earthworm abundance: A quantitative approach. *Journal of Applied Sciences Research*, **3**, 1369-1376.

Kazmierczak, K. (2000). *A Field Guide to the Birds of India, Sri Lanka, Pakistan, Nepal, Bhutan, Bangladesh and the Maldives*. Om Book Service, New Delhi, 352.

Kazmierczak, K. and Singh, R. (1998). *A Birdwatchers' Guide to India*. (eds.: Kazmierczak, K. and Singh, R.) Prion Ltd, Sandy, 82-102.

Khacher, L. (1986). Duck migration across the Himalaya-Tufted Duck, *Aythya fuligula* at 13700m on Rohtang Pass, Himachal Pradesh. *Journal of the Bombay Natural History Society*, **83**, 199-200.

Khaire, N. (2006). *A Guide to the Snakes of Maharashtra, Goa and Karnataka*. Indian Herpetological Society, Pune, 129.

Khanna, V. (2005). Chilopoda: Scolopendridae: *Fauna of Western Himalaya (Part-2)*. Zoological Survey of India Kolkatta, 221-230.

Khatri, T.C. (1997). Butterflies of Car Nicobar. *Indian Journal of Forestry*, **20**, 244-247.

Khera, S. Jindal, R. and Vasisht, H.S. (1984). Avian fauna of Chandigarh (India) and the adjoining areas. *Research Bulletin of the Panjab University*, **35**, 167-175.

Kirby, W.F. (1914). *The fauna of British India including Ceylon and Burma*. Orthoptera (Acrididae). Taylor and Francis Ltd. London, p 276.

Kriplani, M. (1952). On Indian tadpoles wirh suctorial disc. *Records of Indian Museum*, **50**, 359-366.

Krishnamurthy, K.V. (2003). *An advanced text book on Biodiversity. Principles and Practice.* Oxford and IBH Publishing Co. Pvt. Ltd. New Delhi, p 4.

Krombein, K.V. (1982). Biosystematic study of Ceylonese Wasps, IX. A monograph of Tiphiidae (Hymenoptera: Vespoidae). *Smithsonian Contr. Zool*, **374**, 64.

Kulkarni, P.P. and Prasad, M. (2002). Insecta: Odonata. In: *Fauna of Ujani*. Zoological Survey of India, 91-104.

Kumar, A. (1978). Some field notes on Odonata around a fresh water lake in Western Himalayas (Renuka, Himachal Pradesh). *Journal of Bombay Natural History society*, **74**, 506-510.

Kumar, A. (1982). An annotated list of Odonata of Himachal Pradesh. *Indian Journal of Physical and Natural Sciences*, **2**, 55-59.

Kumar, A. (1995). Odonata. In: *Fauna of Western Himalaya-I*. Zoological Survey of India, Kalkata, 25-33.

Kumar, A. (2005). Odonata. In: *Fauna of Western Himalaya (Part 2)*. Zoological Survey of India, Kolkata, 75-98.

Kumar, A. (2010). Hydrological conditions of River Beas and its fish fauna in Kullu Valley, Himachal Pradesh, India. *Environment Conservation Journal*, **11**, 7-10.

Kumar, A. and Juneja, D.P. (1976). The Odonata of Renuka Lake (Western Himalaya: Himachal Pradesh). *Newsletter of Zoological Survey of India*, **2**, 95-96.

Kumar, A. and Prasad, M. (1981). Field ecology, zoogeography and taxonomy of the Odonata of Western Himalaya, India. *Records of Zoological Survey of India, Occasional Paper No.* **20**, 1-118.

Kumar, A., Sati, J.P., Tak, P.C. and Alfred, J.R.B. (2005). *Handbook on Indian Wetland Birds and their Conservation*. Zoological Survey of India, Kolkata, 468.

Kumar, U. and Asija, M. (2006). *Biodiversity: principles and conservation*. Agrobios, Jodhpur, 234.

Kurahashi, H. (1970). The tribe Calliphorini from Australian and Oriental Regions. I. Melinda-group (Diptera: Calliphoridae). *Pacific Insects*, **12**, 519-542.

Kurahashi, H. (1972). A new species of the genus Tricycleopsis from Japan (Diptera: Calliphoridae). *Kontyu*, **40**, 24-26.

Lagler, K.F., John, E.B. and Robert, R.M. (1962). *Ichthyology, the study of fishes*. John Willy and Sons, New York and London, p 543.

Lal, K. (2010). Biology and ecological studies on Orthopteran fauna of Himachal Pradesh. *Ph.D. Dissertation, Himachal Pradesh University*, Shimla, p 213.

Larsen, T. B. (1987). The butterflies of the Nilgiri mountains of Southern India (Lepidoptera: Rhopalocera). *Journal of Bombay Natural History society*, **84**, 26-54.

Lavkumar, K.S. (1962). Bird watching in the Himalayas. *Newsletter for Birdwatchers*, **2**, 6-9.

Lefroy, H.M. and Howlett, F.M. (1909). *Indian Insect Life*. Thacker, Spink and Co., Calcutta, p 786.

Ludlow, F. (1920). Notes on the nidification of certain birds in Ladakh. *Journal of Bombay Natural History society*, **27**, 141-146.

Mac Kinnon J. and Philips, K. (1993). *The Birds of Borneo, Sumatra, Java and Bali*. Oxford University Press, Oxford.

Mackinnon, P.W. and de Niceville, L. (1897). A list of the butterflies of Mussoorie. *Journal of Bombay Natural History society*, **11**, 205-221.

Mackinnon, P.W. and de Niceville, L. (1898). A list of the butterflies of Mussoorie. *Journal of Bombay Natural History society*, 205-221, 368-389, 585-605.

Madhyastha, N. A., Mavinkuruve, R. G. and Shanbhag, S. P. (2004). Land snails of Western Ghats. In A. K. Gupta, A. Kumar and V. Ramakantha (eds), *ENVIS Bulletin*, **4**, 143–151.

Mahabal, A. (1992 a). Avifauna of Chamba District (Himachal Pradesh) with emphasis on their altitudinal distribution. *Pavo*, **30**, 17-25.

Mahabal, A. (1992 b). Natural distribution of some bird species in Chamba District, Himachal Pradesh. *Newsletter for Birdwatchers*, **32** (5-6), 16.

Mahabal, A. (1996). Bird survey in Shiwalik Himalaya of Himachal Pradesh. *Pavo*, **34**, 7-16.

Mahabal, A. (2000 b). Avifauna. In: *Fauna of Renuka Wetland*. Zoological Survey of India, Kolkata, 169-176.

Mahabal, A. (2005). Aves. In: *Fauna of Western Himalaya (Part 2)*. Zoological Survey of India, Kolkata, 275-339.

Mahabal, A. and Mukherjee, R. (1991). Birds of Mandi District (Himachal Pradesh). *Newsletter for Birdwatchers*, **31**, 8-9.

Mahabal, A. and Sharma, T.R. (1992). Distribution patterns of birds of Kangra Valley (Himachal Pradesh). *Himalayan Journal of Environment and Zoology*, **6**, 85-96.

Mahabal, A. and Sharma, T.R. (1993). Birds in Nainadevi Wildlife Sanctuary in Shiwalik Himalayas. *Newsletter for Birdwatchers*, **33**, 43-44.

Mahajan, K.K. and Mukherjee, R. (1974). Brief note on some observation at Lahaul and Spiti, H.P. *Newsletter for Birdwatchers*, **14**, 3-4.

Manakadan, R. and Pittie, A. (2001). Standardised common and scientific names of the birds of the Indian subcontinent. *Buceros*, **6**, 1-37.

Mandal, D.K. and Maulik, D.R. (1991). Insecta: Lepidoptera: Rhopalocera: Nymphalidae: Danainae. In: *Fauna of Orissa*. Zoological Survey of India, Kolkata, 253-238.

Mandal, D.K. and Maulik, D.R. (1998). Lepidoptera: Papilionoidea: Papilionidae and Danaine: Nymphalidae. In: *Fauna of Meghalaya*. Zoological Survey of India, Kolkata, 223-242.

Mandal, D.K., Bhattacharya, D.P., Maulik D.R. and Majumdar, M. (1997). Lepidoptera: Papilionidae and Hesperioidea. In: *Fauna of Delhi*. Zoological Survey of India, Kolkata, 393-407.

Mandal, D.K., Ghosh, S.K. and Majumdar, M. (2000). Insect: Lepidoptera. In: *Fauna of Tripura*. Zoological Survey of India, Kolkata, 283-334.

Mani, A. (1981). *The Himalayan aspects of change*. India International Centre, New Delhi.

Mani, M.S. (1974). *Ecology and Biogeography in India*. Dr. W. Junk, B.V. Publishers, The Hague, Netherland, p 773.

Manjrekar, N. and Mehta, P. (1999). Pond Heron in Pin Valley National Park, Spiti, Himachal Pradesh. *Journal of Bombay Natural History Society*, **96**, 313-314.

Marshall, G. F. L. and de Niceville, L. (1883). *The butterflies of India, Burma and Ceylon*. Taylor and Francis Ltd. London, **1**, 1-321.

Marshall, G. F. L. and de Niceville, L. (1886). *The butterflies of India, Burma and Ceylon*. Taylor and Francis, London, **2**, 1-332.

Marshall, G. F. L. and de Niceville, L. (1890). *The butterflies of India, Burma and Ceylon*. Taylor and Francis Ltd. London, **3**, 1-503.

Marshall, G.A.K. (1916). *The Fauna of British India Including Ceylon and Burma*. Coleoptera: Rhynchophora: Curculionidae (Brachyderinae and Otiorhynchinae); Taylor and Francis Ltd, London, p 366.

Martin, J. W. and Davis, G. E. (2001). An Updated Classification of the Recent Crustacea. Natural History Museum of Los Angeles County, p 132.

Mathew, G. and Rahamathulla, V.K. (1993). Studies on the butterflies of silent valley national park. *Entomon*, **18**, 185-192.

Mattu, V.K. and Thakur, M.L. (2006). Bird Diversity and Status in Summer hill, Shimla (Himachal Pradesh). *Indian Forester*, **132**, 1271-1281.

Mavinkuruve, R. G., Shanbhag, S.P. and Madhyastha, N. A. (2004 a). Checklist of land snails of Karnataka. *Zoos' Print Journal*, **19**, 1684-1686.

Mavinkuruve. R. G., Shanbhag, S.P. and Madhyastha, N. A. (2004 b). Non-marine molluscs of Western Ghats. *Zoos' Print Journal*, **19**, 1708-1711.

McClelland, J. (1839). Indian Cyprinidae. *Asiatic Research*, **19**, 262-450.

McClelland, J. (1842). On the freshwater fishes collection by William Griffith. *Calcutta Journal of Natural History*, **2**, 560-589.

Mehta, H. S. (2009). Fauna of the Indian Cold Desert (Ladakh) -an Overview. ENVIS *Newsletter*, **15**, 2-4.

Mehta, H. S. and Sharma, I. (2008). Pisces. In: *Fauna of Pin Valley National Park*. Zoological survey of India publication, Kolkata, **34**, 75-84.

Mehta, H.S. (2000 a). Pisces. In: *Fauna of Renuka Wetland*. Zoological Survey of India, Kolkata, 141-149.

Mehta, H.S. (2000 b). Amphibia. In: *Fauna of Renuka Wetland*. Zoolological Survey of India, 151-161.

Mehta, H.S. (2000 c). Reptilia, In: *Fauna of Renuka wetland*. Zoological Survey of India, 163-168.

Mehta, H.S. (2005). *Fauna Western Himalaya (Part-2)*. Zoological Survey of India, Kolkata, p 359.

Mehta, H.S. and Julka, J.M. (2002). Mountains: Northwest Himalaya. In: *Ecosystems of India*. (Eds. Alfred, J.R.B., Das, A.K. and Sanyal, A.K.). ENVIS Centre, Zoological Survey of India, Kolkata, p 410.

Mehta, H.S. and Uniyal, D.P. (2005). Pisces. In: *Fauna of Western Himalaya (Part 2)*. Zoological Survey of India, Kolkata, 255-268.

Mehta, H.S., Mattu, V.K. and Thakur, S.K (2002 a). Orthopteran diversity of Kalatop -khajjiar wildlife sanctuary, Chamba, (H.P) India. *Bionotes*, **4**, 60.

Mehta, H.S., Sharma, R.M., Mitra, B. and Thakur, M.S. (2003). A preliminary study on the butterflies of Himachal Pradesh as flower visitors and pollination. *Annals of Entomology*, **21**, 1-3

Mehta, H.S., Thakur, M.S., Sharma, R.M. and Mattu, V.K. (2002 b). Butterflies of Pong Dam wetland, Himachal Pradesh. *Bionotes*, 37-38.

Mehta, H.S.; Thakur, M.L.; Paliwal, R. and Tak, P.C. (2002 c). Avian diversity of Ropar Wetland, Punjab, India. *Annals of forestry*, **10**, 307-326.

Mehta, S.L. (2003). Biosystematic studies on insect fauna in Shimla hills, Himachal Pradesh. *M. Phil. Dissertation, Himachal Pradesh University*, 79-83.

Melchias, G. (2001). *Biodiversity and conservation*. Oxford and IBH Publ. Co., New Delhi, p 236.

Menon, A.G.K. (1951). Note on fishes in the Indian Museum. XLVII. On two new species of the genus *Nemachilus* from Kangra Valley, Punjab. *Indian Museum of science*, **49**, 227-230.

Menon, A.G.K. (1954). Fish geography of Himalayas. *Proceedings of the national institute of sciences, India*, **20**, 467-493.

Menon, A.G.K. (1962). A distribution list of fishes of the Himalayas. *Journal of Zoological Society of India*, **14**, 23-32.

Menon, A.G.K. (1974). *A checklist of fishes of the Himalayan and the Indo-Gangetic plains*. Special publication no. 1. Inland Fisheries Society of India, Barrackpore, India, p 136.

Menon, A.G.K. (1987). Pisces, 4 (I). In: *Fauna of India and the Adjacent Countries.* Zoological Survey of India, Kolkata, p 259.

Menon, A.G.K. (1999). Checklist of freshwater fishes of India. *Records of Zoological Survey of India*, **175**, 366.

Mistry, N.M. (1967). Bird watching on a Simla-Kullu trek. *Newsletter for Birdwatchers*, **7**, 2-4.

Mizukami, R.N., Goto, M., Izumiyama, S., Hayashi, H. and Yoh, M. (2005). Estimation of feeding history by measuring carbon and nitrogen stable isotope ratios in hair of Asiatic black bears. *Ursus*, **16**, 93-101.

Mondal, D.K. (1998). Lepidoptera. In: *Faunal Diversity of India*. Zoological Survey of India Publication, Kolkata, 311-318

Morely, C. (1913). *The Fauna of British India including Ceylon and Burma*. Hymenoptera-III. Taylor and Francis Ltd. London, p 531.

Mukherjee, R.N. and Chandra, M. (1984). Birds of Sili Forest, Solan, Himachal Pradesh. *Newsletter for Birdwatchers*, **24**, 14-15.

Mukhopadhyay, P. and Sharma, R. M. (2008). Insecta: Coleoptera. In: *Fauna of Pin Valley National Park*. Zoological Survey of India publication, Kolkata, **34**, 85-88.

Myers, N., Mittermeier, R.A., Mittermeier, C.G., Da Fonseca, G.A.B. and Kent, J. (2000). Biodiversity hotspots for conservation priorities. *Nature*, **403**, 853-858.

Najar, I.A. and B. Khan, A.B. (2011). Earthworm communities of Kashmir Valley, India. *Tropical Ecology*, **52**, 151-162.

Nameer, P.O., Unnikrishnan, K.R. and Thomas, J. (2007). Record of Leith's Softshell Turtle *Aspideretes leithii* (Gray, 1872) (Family Trionychidae) from Kerala. *Reptile Rap*, **8**, 5.

Nandi, B. C. (2002 b). Diptera (Sarcophagidae). In: *Fauna of India*. Zoological Survey of India, Kolkata, p 608.

Nandi, B.C. (2002 a). Blow Flies (Diptera: Calliphoridae) of West Bengal, India with a note on their biodiversity. *Records of Zoological Survey of India*, Kolkata, **100**, 117-129.

Nandi, B.C. (2004). Checklist of Calliphoridae (Diptera) of India. *Records of zoological Survey of India*, **231**, 1-47.

Nandi, D.N. (1987). Insecta: Lepidoptera: Rhopalocera: Nymphalidae. In: *Fauna of Orissa*. Zoological Survey of India, Kolkata, 193-203.

Nandi, D.N. (1993). Insect: Lepidoptera: Satyridae. In: *Fauna of Orissa*. Zoological Survey of India, Kolkata, 181-184.

Narang, M.L. (1986). Contribution to food habits of Common Babbler, *Turdoides caudatus* (Dumont). *Indian Journal of Forestry*, **9**, 140-145.

Narang, M.L. (1989). Birds of Sangla Valley. *Newsletter for Birdwatchers*, **29**, 8.

Narang, M.L. and Singh, A.P. (1995). Birds of Nauni campus of University of Horticulture and Forestry, Solan, Himachal Pradesh. *Newsletter for Birdwatchers*, **35**, 106-108.

Narwade, S.S., Jathar, G.A. and Rahmani, A.R. (2006). Himachal Pradesh. In: Bibliography of the birds of North India. *Buceros*, **11**, 34-54.

Negi, S.S. (2005). *Himalayan Wildlife Habitat and Conservation*. Indus Publishing Company, **47**, 40-63.

Nelson, J.S. (2006). *Fishes of the World*. Fourth Edition, John Wiley and Sons, Inc., 1-601.

Neumann, P. and Elzen, P.J. (2004). The biology of the small hive beetle (Aethina tumida Murray, Coleoptera: Nitidulidae): Gaps in our knowledge of an invasive species. *Apidologie*, **35**, 229–247.

Nevill, G. (1878). Mollusca II. Mollusca from Kashmir and neighbourhood of Mari (Murree) in the Punjab. *Scientific Records of Second Yarkand Mission*. Mollusca, London, 14-21.

Norse, E.A. and McManus, R.E. (1980). Ecology and living resources biological diversity. In: *Environmental Quality 1980*. 11th Annual report of the council on environmental quality, Washington DC, 31-80.

Noss, R. (1990). Indicators for Monitoring Biodiversity: A Hierarchial Approach. *Conservation Biology*, **4**, 355-364.

Nowak, R. (1999). *Walker's Mammals of the World*. Johns Hopkins Univ. Press, Baltimore, Maryland.

Osmaston, B.B. (1926). The birds of Ladakh. *The International Journal of Avian Science (Ibis)*, **2**, 446-448.

Pai, I.K. (2002). Butterfly distribution pattern in Goa. *Insect Environment*, **7**, 165-167.

Pajni, H.R. and Gera, H.C. (1983). A report on the Muscidae and Colliphoridae oc Chandigarh and the surrounding areas (Diptera: Calyptrata). *Research Bulletin of the Panjab University, Science*, **34**, 7-13.

Paliwal, R. (1994). Taxonomy and ecology of earthworms of western Himalaya. *Ph.D. thesis, Ch. Charan Singh University*, Meerut (U.P.), 364.

Paliwal, R. (2008). Annelida: Oligochaeta. In: *Fauna of Pin Valley National* Park. Zoological Survey of India, Kolkata, **34**, 11-22.

Pandey, S. (1989 a). Some observations on the birds of Pin Valley National Park. *Newsletter for Birdwatchers*, **29**, 9.

Pandey, S. (1989 b). The birds of Pong Dam Lake Bird Sanctuary. *Tigerpaper*, **16**, 20-26.

Pandey, S., Sathyakumar, S. and Thakur, M.L. (2004). Kalatop Khajjiar Wildlife Sanctuary. In- *Important Bird Areas in India: Priority Sites for Conservation*. (eds.: Islam, M.Z. and Rahmani, A.R.) *Indian Bird Conservation Network*. BNHS & Birdlife International (UK), 445-446.

Parui, P. and Mukherjee, M. (2000). Diptera. In: *Fauna of Renuka Wetland*. Zoological Survey of India, Kolkata, 135-140.

Parui, P., Mitra, B. and Sharma, R.M. (2006). Diptera Fauna of Punjab and Himachal Shiwalik Hills. *Records of zoological Survey of India*, **106**, 83-108.

Patil, S.G. (2008). Mollusca: *Fauna of Pin Valley National Park (Himachal Pradesh)*. Zoological Survey of India, Kolkata, 25-28.

Peck, S.V. (2006). Distribution and biology of the ectoparasitic beaver beetle Platypsyllus Castoris Ritsema in North America (Coleoptera: Leiodidae: Platypsyllinae). *Insecta Mundi*, **20**, 85–94.

Peterson, H. B. (1915). *The Fauna of British India- Including Ceylon and Burma*. Mollusca III. Taylor and Francis Ltd. London, 244.

Pomeroy, D. and Tengecho, B. (1986). Studies of birds in a semi-arid area of Kenya. III- the use of 'timed species-counts' for studying regional avifauna. *Journal of Tropical Ecology*, **2**, 231-247.

Ponniah, A.G. and Gopalakrishnan, A. (2000). *Endemic Fish Diversity of the Western Ghats*. NBFGR- NATP Publication, **1**, 1-347.

Ponniah, A.G. and Sarkar, U.K. (2000). *Fish Biodiversity of North-East India*. NBFGR-NATP Publication, **2**, 1-228.

Poorni, J. (2003). A new species of *Telsimia* Casey (Coleoptera: Coccinellidae) predatory on arecanut scale from Karnataka, India. *Entomon*, **28**, 51-53.

Prasad, M and Kulkarni, P.P. (2001). Insecta: Odonata. In: *Fauna of Nilgiri Biosphere Reserve*. Zoological Survey of India, 73-83.

Prasad, M and Kulkarni, P.P. (2002). Insecta: Odonata. In: *Fauna of Eravikulam national park*. Zoological Survey of India, **13**, 7-9.

Prasad, M. (1996). An account of Odonata of Maharashtra state, India. *Records of Zoological Survey of India*, **95**, 305-327.

Prasad, M. (1998). *Odonata*. In: *Faunal Diversity in India*. Zoological Survey of India, Kolkata, 172-178.

Prasad, M. and Varshney, R.K. (1995). A Checklist of the Odonata of India including data on larval studies. *Oriental insects*, **29**, 385-428.

Prashad, B. (1914). Lizards of the Shimla-Hill States. *Records of Indian Museum*, **10**, 367-369.

Prater, S.H. (1980). *The book of Indian animals*. III edition. Bombay Natural History Society, Bombay. Oxford University Press, p 324.

Price, T. and Jamdar, N. (1990). The breeding birds of Overa Wildlife Sanctuary, Kashmir. *Journal of Bombay Natural History Society*, **87**, 1-15.

Radhakrishnan, C. and Sharma, R.M. (2002). Butterflies. In: *Fauna of Eravikulum national park*. Zoological Survey of India, Kolkata, 35-40.

Rahbek, C. and Graves, G.R. (2001). Multiscale assessment of patterns of avian species richness. *Proceedings of National Academy of Sciences, USA.*, **98**, 4534-4539.

Rajagopal, A.S. and Subba Rao, N.V. (1968). Aquatic and amphibian molluscs of the Kashmir valley, India. *Proceedings of Symposium on Mollusca*, Part-I, 95-120.

Rajagopal, A.S. and Subba Rao, N.V. (1972). Some land molluscs from the Kangra valley. *Records of Zoological Survey of India*, **66**, 197-212.

Ramakrishna and Mitra, S. C. (2002). Endemic land molluscs of India. *Records of Zoological Survey of India*, **196**, 1-65.

Ramakrishna, G. (1995). *The Fauna of India and the Adjacent Countries.* Zoological Survey of India, Kolkata, 1-130.

Rana, B.S. (1997). A record of Pallas' Fishing Eagle, *Haliaeetus leucoryphus* from Spiti Valley (H.P.). *Journal of Bombay Natural History society*, **94**, 400.

Rao, K.V. Surya and Mitra, S.C. (2005). Mollusca. In: *Fauna of Western Himalaya (Part 2)*. Zoological Survey of India, Kolkata, 39-51.

Rao, N.V. Subba (1998). Mollusca. In: *Faunal Diversity in India.* Zoological Survey of India, Kolkata, 103-117.

Reeves, S.K. (1981). Birds of Sukhna lake. *Newsletter for Birdwatchers*, **21**, 10-11.

Rizvi, A.N. (2010). Soil Nematode Fauna of Ladakh. *Envis Newsletter*, Zoological Survey of India, **16**, 15-16.

Roberts, T.J. (1977) *The mammals of Pakistan*. Ernst Benn Ltd., London and Tonbridge, p 345.

Robertson, I. A. D. (1967). Field records of saltatorial Orthoptera collected in western Tanzania. *Proceedings of the Royal entomological Society*, London (A), **42**, 1-17.

Rodgers, W.A. and Panwar, S.H. (1988). Planning a Wildlife Protected Area Network in India. Vols. I and II. Wildlife Institute of India, Dehra Dun.

Rodgers, W.A., Panwar, H.S. and Mathew, V.B. (2002). Wildlife Protected Areas Network in India: A Review (executive summary). Wildlife Institute of India, Kolkatta. p 43.

Ronges, K. (2009). Revision of Oriental species of *Bengalia peuhi* species group (Diptera, Calliphoridae). *Zootaxa*, **2251**, 1-76.

Rouse, G.W. (2002). *"Annelida (Segmented Worms)"*. Encyclopedia of Life Sciences. John Wiley & Sons, Ltd., doi:10.1038/npg.els.0001599.

Ruppert, E.E., Fox, R.S., and Barnes, R.D. (2004). Annelida. In: *Invertebrate Zoology* (7ed.). Brooks / Cole. 414–420.

Saha, S.S. (1998). Aves. In: *Faunal Diversity of India*. Envis Centre, Zoological Survey of India, Kolkata, 449-457.

Saikia U. and Mehta H.S. (2009). *Faunal Biodiversity of Pong* Dam. Zoological Survey of India, **12**, 99-108.

Saikia U. and Sharma D.K. (2009). *Faunal Biodiversity of Simbalbara wildlife* Sanctuary. Zoological Survey of India, Kolkata, **41**, 65-79.

Saikia, U. and Sharma, D. K. (2008). Reptilia. In: *Faunal Diversity of Simbalbara WLS.* Zoological Survey of India, Kolkata, **41**, 65-79.

Saikia, U. and Sharma, D.K. (2008). Herpetofauna. In: *fauna of Pin Valley National Park.* Zoological Survey of India, 34, 93-96.

Saikia, U., Mehta, H. S. and Sharma, D. K. (2010 b). New distributional record of Eastern Black Turtle Melanochelys trijuga indopeninsularis from Simbalbara WLS, Sirmour district,H. P. *The Indian Forester*, **136**, 273-275.

Saikia, U., Sharma, D. K. and Mehta, H. S. (2010 c). First record of large worm snake (Typhlops diardii) in Himachal Prdaesh. *The Indian Forester*, **163**, 53-56.

Saikia, U., Sharma, D.K. and Sharma, R.M. (2007). Checklist of the Reptilian fauna of Himachal Pradesh. *Reptile Rap*, **8**, 6-9.

Saikia, U., Sharma, N. and Dass, A. (2010a). Vanishing species: The planet in crisis. *Resonanace*, 321- 336.

Saikia, U., Thakur, M. L., Bawri, M. and Bhattacharjee, P.C. (2011). An inventory of the Chiropteran fauna of Himachal Pradesh, northwestern India with some ecological observations. *Journal of Threatened Taxa*, **3**, 1637-1655.

Saini, K. and Mehta, H.S. (2007). An inventory of the Orthoptera insects of Himachal Pradesh. *Bionotes*, **9**, 76-78.

Saini, M.S. and Ghattor, S. (2007). Taxonomy and food plants of some Bumble bee species of lahul and Spiti valley of Himachal Pradesh. *Zoos' Print Journal*, **22**, 1648-2657.

Sangha, H.S. (2005). New and significant records from the Great Himalayan National Park, Himachal Pradesh, India. *Indian Birds*, **1**, 33-34.

Santharam, V. (2005). Birds seen on a trek in the Chansal Pass, Himachal Pradesh. *Indian Birds*, **1**, 28-31.

Saraswat, S. (2002). Insect diversity studies of Shimla hills, Himachal Pradesh. *M. Phil. Dissertation, Himachal Pradesh University*, 81-83.

Sathianarayanan, A. and Khan, A.B. (2006). Diversity, distribution and abundance of earthworms in Pondicherry region. *Tropical Ecology*, **47**, 139-144.

Saussure, H.De. (1884). Prodromus *Oedipodiorum insectorum* ex Ordina Orthopterorum. *Memoires de la Societe du museum d'histoire naturelle de Geneva*, **28**, 1-254.

Saussure, H.De. (1888). Addimenta ad prodomum, *Oedipodiorum insectorum* ex Ordine Orthopterorum. Mémoires de la Socièté de Physique et d'Histoire Naturelle de Genève, **30**, 1-180.

Seddon, M. (2000). Molluscan Biodiversity and the Impact of Large Dams. Report to IUCN Species Survival Commission.

Sengupta, T. and Pal, T.K. (1998). Faunal Diversity in India: Coleoptera. *Records of Zoological Survey of India, Kolkata*, 132-139.

Sharma, D. K., Paliwal, R. and Saikia, U. (2009). Aves: *Faunal Diversity of Simbalbara Wildlife Sanctuary*. Zoological Survey of India Publication, **41**, 81-101.

Sharma, D.K. (2005). Record of Leith's Sand Snake from Himachal Pradesh. *Bionotes,* **7**, 53.

Sharma, D.K. and Saikia U. (2009). Mammals: *Faunal Diversity of Simbalbara Wildlife Sanctuary*. Zoological Survey of India Kolkata, **4**, 103-118.

Sharma, I. and Mehta, H.S. (2009). Fishes of Ladakh. *ENVIS News Letter*, **15**, p 13.

Sharma, K.B., Anumehra and Khan, K.M. (2006). Feeding Behaviour of Black Buck (Antelope cervicapra) In Gangetic Plains of Bihar. In: *Biodiversity and Environment* (Eds.: Pandey B.N. and Kulkarni G.K.). A.P.H. Publication, New Delhi 171-180.

Sharma, R.M. (2008). *Fauna of Pin Valley National Park*. Zoological Survey of India Kolkatta , **34**, 1-10.

Sharma, R.M., Bulganian M. and Maheshwary G. (2004).Beetles of Kalatop-Khaijjair wildlife sanctuary, Himachal Pradesh. *Zoos' Print Journal*, **19**, 1626.

Sharma, T.R. and Mahabal, A. (1997). Seasonal changes of bird species in two different altitudinal locations of Solan District, Himachal Pradesh. *Records of the Zoological Survey of India*, **96**, 151-166.

Sharma, V.K. and Tandon, K.K. (1990). The fish and fisheries of Himachal Pradesh state of India. *Publication of Fish Bulletin*, **14**, 41-46.

Shinde, K. and Sathe, T.V. (2006). Biodiversity of dragonflies (Odonata) from koyna dam and around area. In: *Biodiversity and Environment* (Eds: Pandey B.N. and Kulkarni G.K.). A.P.H. Pub., New Delhi, 61-65.

Shishodia, M.S. and Gupta, S. (2010). Checklist of Orthoptera (Insecta) of Himachal Pradesh, India. *Journal of Threatened Taxa*, **1**, 569-572.

Shishodia, M.S. and Tandon S.K. (1993). Insecta: Orthoptera: Grylloidea. In: *Fauna of West Bengal*. Zoological Survey of India, **4**, 227-285.

Shishodia, M.S. and Tandon, S.K. (2000). Orthoptera. In: *Fauna of Renuka Wetland*. In: Zoological Survey of India, 73-90.

Shishodia, M.S., Mehta, H.S. Mattu. V.K. and Thakur, S.K. (2002). Orthoptera (Insecta) Pong dam wetland. Distt. Kangra (H.P.) India. *Zoos' Print Journal*, **18**, 1047-48.

Sibley, C.G. and Monroe, B.L. (1990). *Distribution and Taxonomy of Birds of the World*. Yale Univ. Press, New Haven, Connecticut.

Sidhu, I.S. and Singh. D. (2002). Blowflies (Diptera: Calliphoridae) collected from North-Western region including two new records from India. *U. P. J. Zool*, **22**, 93-95.

Singh, A.P. (2003). Birds of Tabo: a lesser known cold desert in the Western Himalaya. *Journal of Bombay Natural History Society*, **100**, 152-154.

Singh, D. and Sidhu, I.S. (2004a) A checklist of blow flies (Diptera: Calliphoridae) from North-West of India. *Uttar Pradesh Journal of Zoology*, **24**, 63-71.

Singh, D. and Sidhu, I.S. (2004b). New records of blow flies (Diptera: Calliphoridae) from India. *Entomon*, **29**, 203-206.

Singh, D. and Sidhu, I.S. (2007). Two New species of Melinda Robineau-Desvoidy (Diptera: Calliphoridae) from India, with a key to the Indian species of this genus. *Journal of Bombay Natural History Society*, **104**, 55-57.

Singh, M. (2001). Insect diversity studies of Sukhna and Catchment area. *M.Phil. Dissertation, Himachal Pradesh University*, 84-85.

Singh, R. (2007). Studies on insect fauna of chandertal wetland of Lahaul and Spiti district (H.P.). *M.Phil. Dissertation, Himachal Pradesh University, Shimla*, 47-57.

Singh, R. (2010). Studies on faunal diversity of Chandertal wildlife sanctuary in Lahaul and Spiti district of Himachal Pradesh. *Ph.D. Dissertation, Himachal Pradesh University*, Shimla, p 197.

Singh, S. (1963). Entomological survey of Himalayas, Part- XXVI, fourth and final annotated check list of insect from North- West (Punjab) Himalayas. *Agra University Journal of Research*, **12**, 363-393.

Singh, S. and Maheshwari, G. (1986). Chironomidae (Diptera) as indicator of lake typology of North West Himalayas, India. *Advances in Limnology*, 45-56.

Singh, S. and Maheshwari, G. (1987). Chironomidae (Diptera) of Chandertal Lake Lahul Valley (NW Himalayas). *Annals of Entomology*, **5**, 11-20.

Singh, S., Kothari, A. and Pande, P. (1990). Directory of National Parks and Sanctuaries in Himachal Pradesh: Management, Status and Profiles. *Indian Institute of Public Administration,* Environmental Studies Division, New Delhi.

Singh, V. and Banyal, H.S (2012). Diversity and Ecology of Mammals in Kalatop-Khajjiar Wildlife Sanctuary, District Chamba (Himachal Pradesh), India. International Journal of Science and Nature, 3, 125-128.

Singh, V. and Banyal, H.S (2014). First Record of Kashmir Rock Agama (Laudakia tuberculata Hardwicke & Gray, 1827) from Kalatop-Khajjiar Wildlife Sanctuary, Chamba (Himachal Pradesh), India. Russian Journal of Herpetology, **21**, 234 – 236

Singh, V.J. (2001). Harike: Wetland haven. *Sanctuary Asia*, **21**, 30-35.

Sinha, R.K. (2009). Earthworms: the miracle of nature (Charles Darwin's'unheralded soldiers of mankind & farmer's friends'). *Environmentalist*, **29**, 339-340.

Smith, M. (1935). *The Fauna of British India including Ceylon and Burma*. Reptilia and Amphibia. Vol. **2**.Sauria. Taylor and Francis Ltd., London, p 583.

Smith, M. (1943). *The Fauna ofBritish India, Ceylon and Burma Including the Whole of the Indo-Chinese Sub Region.* Reptilia and Amphibia. Vol. III. Taylor and Francis Ltd. London, p 583.

Snedecore, G.W. and Cochran, W.G. (1993). *Statistical Methods*. Oxford and IBH Publ. Co., New Delhi.

Solbrig, O.T. (2000). The theory and practice of the science of Biodiversity: A personal assessment. In: *Biology of Biodiversity*. (Ed.: Kato, M). Springer Hong Kong, 107-117.

Southwick, C.H. and Siddiqi, M.F. (1994) Primate commensalisms: the rhesus monkey in India . *Rev Ecol (Terre Vie)*, **49**, 223-31

Stal, C. (1860). Orthoptera. Species novas descripsit. *Eugenic's Resa, Orth., Stockholm*, **3**, 299-350.

Stal, C. (1873). Revue critique des Orthopteres descripstes per Linne. *Der Deo Geeret Thunberg. Recens. Orth*, **1**, 154.

Stattersfield, A.J., Crosby, M.J., Long, M.J. and Wege, D.C. (1998). *Endemic Bird Areas of the World: Priorities for Biodiversity Conservation*. (Conservation Series 7) BirdLife International, Cambridge, UK.

Stebbing, E.P. (1914). *Indian Forest Insects of Economic Importance*. Eyre and Spottiswoode, Ltd., London.

Stenidachner, F. (1867). Ichthyologische Notizen. IV.Sitzungsberichte der Konigl Akademie der Wissenschaften zu Munchen, **55**, 517-534.

Stephenson, J. (1923). *The Fauna of British India including Ceylon and Burma*. Oligochaeta. Taylor and Francis Ltd., London, 518 p.

Stoliczka, F. (1868). Ornithological observations in the Sutlej Valley, Northwest Himalayas. *Journal of Asiatic Society of Bengal*, **37**, 1-70.

Stork, N.E. (1988). Insect diversity: facts, fiction and spaculation. *Biological Journal of Linnean Society*, **35**, 321-337.

Subba Rao, N.V. and Mitra, S.C. (1995). *Mollusca Fauna of Western Himalaya (UP)*. Zoological Survey of India, **1**, 11-15.

Sureshan, P.M. (2009). New findings of Hymenoptera (Insecta) from Ladakh. *ENVIS Newsletter*, **15**, 10-12.

Suyal, B.O. (1992). Birds of Sarahan Bushar, Shimla District. *Newsletter for Birdwatchers*, **32**, 14-15.

Swinhoe, C. (1909-1913). Lepidoptera Indica.Vol. **VII**-X

Tak, P.C. (1987). On a rare sighting of Western Tragopan (*Tragopan melanocephalus*) in district Chamba, Himachal Pradesh, India. *Cheetal*, **28**, 42-45.

Takele, S., Bekele, A., Gurja, B. and Balakrishnan, M. (2011). A comparison of rodent and insectivore communities between sugarcane plantation and natural habitat in Ethiopia. *Tropical Ecology*, **52**, 61-68.

Talwar, P.K. and Jhingram, A.G. (1991). *Inland fishes of India and adjacent countries*. Oxford publication, New Delhi, **1-2**, 1-1158.

Tandon, S.K. and Shishodia, M.S. (1969). On a collection of Acridoidea (Orthoptera) from the Nagarjuna Sagar dam area. *Oriental Insects*, **3**, 395-402.

Tandon, S.K. and Shishodia, M.S. (1995). Insecta: Orthoptera. In: *Fauna of Western Himalaya, Part*-1. Zoological Survey of India, Kolkota, 37-42.

Tandon. S.K. (1976). A check-list of the Acridoidea (Orthoptera) of India Part I Acrididae, *Records of Zoological Survey of India, Kolkata*, **3**, 1-48.

Tandon. S.K. and Khera, S.K. (1978). Ecology and distribution of grasshoppers (Orthoptera : Acridoidea) in Arunanchal Pradesh, India and the impact of human activities on their ecology and distribution. *Mem. School. Entomol.*, **6**, 73-91.

Thakur, M.L. (2008). Studies on status and diversity of avifauna of Himachal Pradesh. *Ph.D. thesis, Himachal Pradesh University, Shimla*, 306.

Thakur, M.L. and Mattu, V.K. (2010). The Role of butterfly as flower visitors and pollinators in Shiwalik Hills of Western Himalayas. *Asian Journal of experimental Biological Sciences*, **1**, 822-825.

Thakur, M.L. and Mattu, V.K. (2011). Avifauna of Kaza area of Spiti (Himachal Pradesh), India. *International Journal of Science and Nature*, **2**, 483-487.

Thakur, M.L. and Mattu, V.K. (2012). *Birds of Himachal Pradesh, India*. LAP-Lambert Academic Publishing GmbH & Co., Saarbrucken, Germany, p 369.

Thakur, M.L. and Paliwal, R. (2012). Avian Diversity of Chandigarh (UT). *International Journal of Advance biological Research*, **2**, 103-114.

Thakur, M.L., Mattu, V.K. and Sharma, R.M. (2006 b). Bird diversity and status in Tara Devi, Shimla, Himachal Pradesh. In: *Biodiversity and Environment* (Eds.: Pandey B.N. and Kulkarni G.K.). A.P.H. Publication, New Delhi.

Thakur, M.L., Mattu, V.K., Hira Lal, Sharma, V., Hem Raj and Thakur, V. (2010 b). Avifauna of Arki Hills, Solan (Himachal Pradesh), India. *Indian Birds*, **5**, 162-166.

Thakur, M.L., Mattu, V.K., Mattu, N., Sharma, V., Bhardwaj, R. and Thakur, V. (2010 a). Birddiversity in Sarkaghat valley, (Himachal Pradesh), India. *Asian Journal of Experimental Biological Sciences*, **1**, 940-950.

Thakur, M.S., Mattu, V.K. and Mehta S. L. (2008). Distributional record of insect diversity in different altitudes of Shimla hills, Himachal Pradesh, India. *Journal of Entomological Research*, **32**, 317-321.

Thakur, M.S., Mattu, V.K. and Mehta, H.S. (2006 a). Studies on the butterflies of Sukhna and catchment area in Chandigarh, India. *Journal of Entomological Research*, **30**, 175-178.

Thakur, S.K. and Mattu, V.K. (2006). Orthopteran diversity of Pin Valley National Park, Lahaul and Spiti, India. *Zoos' Print Journal*, **21**, 2225.

Theobald, W. (1862). Notes on a trip from Simla to the Spiti valley and Chomoriri (Tshomoriri) Lake during the months of July, August and September, 1861. *Journal of Asiatic Society of Bengal*, **31**, 480-527.

Theobald, W. (1878). Notes on the land and freshwater shells of Kashmir, more particularly of the Jhelum and the hills of Jammu, *Journal of Asiatic Society of Bengal*, **47**, 141-149.

Thomas, O. (1915). Scientific results from the mammal survey No. 10: The Indian bats assigned to the genus *Myotis*. *Journal of the Bombay Natural History Society*, **23**, 607–612.

Tilak, R. and Husain, H. (1977). A checklist of fishes of Himachal Pradesh. *Zool. Jb. Syst. Bd.*, **104**, 265-301.

Tilak, R. and Mehta, H.S. (1983). On a collection of amphibians of the Sirmour District (Himachal Pradesh). *Research Bulletin of Panjab University*, **34**, 157-166.

Tinkham, E.R. (1937). Spathosternum sinese Uvaroov, considered to be a race of *Spathosternum prasiniferum* (Walker) Orthoptera: Acrididae, *Lingn. Science Journal, Canton*, **15**, 47-54.

Tiwari, S.K. (2007). Biogeography of India. *Geographic Biogeography*, 159-251.

Tiwari, T. (2011). Chemoattractive Effect of Amino AcidsAgainst *Lymnaea acuminata* Snails. *World Journal of Zoology*, **6**, 117-119.

Tripathi, G. and Bhardwaj, P. (2004). Earthworm diversity and habitat preferences in arid regions of Rajasthan. *Zoo's Print Journal*, **19**, 1515-1519.

Tytler, R.C. (1868). Notes on the birds observed during a march from Simla to Mussoorie. *The International Journal of Avian Science Ibis*, **4**, 190-203.

UNEP (1995). *Global biodiversity assessment*. Cambridge University Press, Cambridge.

Uvarov, B.P. (1925). Insects, Orthoptera, Acriididae. In: mission Guy Babault dens I inde et drans la region occidentale-I. *Himalaya*, 1914 Paris, 40.

Uvarov, B.P. (1926). Notes on the genus *Oxya*. *Bulletin of Entomological Research, London*, **17**, 45-48.

Uvarov, B.P. (1928 a). Distributional records of Indian Acrididae. *Records of Indian Museum*, **29**, 233-239.

Uvarov, B.P. (1928 b). *Locusts and grasshoppers*. A hand book for their study and control London, p 352.

Uvarov, B.P. (1929). Acrididen (Orthoptera) aus Sud Indian. *Rev. Suisse Zoology, Geneva*, **36**, 533-563.

Varshney, R.K. (1992). A check list of scale insects and Mealy bugs of Southe Asia. Part-I. *Records of Zoological Survey of India*, **139**, 1-152.

Varshney, R.K. (1993). Index Rhopalocera Indica. Part III. Genera of butterflies from India and neighbouring countries (Lepidoptera: (A) Papilionidae, Pieridae and Donaidae). *Oriental Insects*, **27**, 347-372.

Varughese, G.C., Vijyalakhhmi, K., Kumar, A. and Rana, N. (2009). *State of environment report, India*. ENVIS, Ministry of Environment and Forestry, Government of India, 1-73.

Vasudevan, K., Kumar, Ajith & Chellam, Ravi. (2001). Structure and composition of rain forest amphibian communities in Kalalad-Mundunthurai Tiger Reserve. *Current Science*, **80**, 406-412.

Vats R. and Gupta S.K. (2011). Icthyofauna of four district of northern Haryana. *Journal of Arts, Science and Commerce*, **4**, 1-7.

Vedwan, N. and Rhodes, R.E. (2001). Climate change in the Western Himalayas of India: a study of local perception and response. *Climate Research*, **19**, 109-117.

Venkataraman, K. and Krishnamoorthy (1998). Crustacea. In: *Faunal Diversity in India*. Zoological Survey of India, Kolkata, 134-143.

Venugopal, P.D. (2010). An updated and annotated list of Indian lizards (Reptilia: Sauria) based on a review of distribution records and checklists of Indian reptiles. *Journal of Threatened Taxa*, **2**, 725-738.

Verma, R.K. and Kapoor, K.S., 2011. Kalatop Khajjiar wildlife sanctuary Chamba, Himachal Pradesh: An appraisal to its Plant diversity. HFRI, Shimla. pp.124

Vinitpornsawan, S., Steinmetz, R. and Kanchanasakha, B. (2006). *The status of bears in Thailand*. Japan Bear Network, Ibaraki, Japan.

Vishwanath, W., Lakra, W.S. and Sarkar, U.K. (2007). *Fishes of North East India.* National Bureau of Fish Genetic Resources Publication, 1-264.

Vyas, R. (2007). Herpetofauna of Purna Wildlife Sanctuary, Gujarat, India. *Reptile Rap*, **8**, 10-15.

Walia, V.K. (2005). Insecta: Lepidoptera: Geometridae (Moths). In: *Fauna of Western Himalaya (Part 2)*. Zoological Survey of India, Kolkata, 181-190.

Walker, F. (1869). Characters of some apparently undescribed Ceylon Insects. *Annual Magazine of Natural History*, **4**, 217-224.

Walker, F. (1870). Catalogue of the specimens of *Dermaptera saltatoria*. Collections of British Museum London, **3**, 425-604.

Walker, F. (1871). Catalogue of the specimens of *Dermaptera saltatorial.* Collections of British Museum, **5**, 811-859.

Waltner, R.C. (1974). Geographical and altitudinal distribution of amphibians and reptiles in the Himalayas. *Cheetal*, **16** (1), 17-25; **16** (2), 28-36; **16** (3), 14-19; **16** (4), 12-17.

Wanger, R., Leese, F. and Panesar, A.R. (2004). Aquatic Dance Flies from a small Himalayan Mountain Stream System (Diptera Empididae: Hemerodromiinae, Trichopezinae and Clinoceinae). *Bonner Zoologische Beitrage*, **52**, 3-31.

Westwood, J.O. (1832). Descriptions of several new British forms amongst the parasitic hymenopterous insects. *Philosophical Magazine*, **1**, 127–129.

Whistler, H. (1926 a). The birds of the Kangra District, Punjab, part-1. *The International Journal of Avian Science (Ibis)*, **2**, 521-581.

Whistler, H. (1926 b). A note on the birds of Kullu. *Journal of the Bombay Natural History Society*, **31**, 458-485.

Whittaker, R.H. (1972). Evolution and measurement of species diversity. *Taxon*, **21**, 231-251.

Wilson, D.E. and Reeder, D.M. (1993). *Mammals species of the World. A Taxonomic and Geographic Reference.* Smithsonian Institute Press, Washington and London.

Wilson, E.O. and Peters, F.M. (1988). *Biodiversity*. National Academy Press, Washington DC.

Wynter-Blyth, M.A. (1948). An expedition to Sangla in Kunawar. *Journal of Bombay Natural History Society*, **47**, 565-585.

Wynter-Blyth, M.A. (1951). A naturalist in the north-west Himalaya. *Journal of Bombay Natural History Society*, **50**, 559-572.

Wynter-Blyth, M.A. (1952). A naturalist in the Northwest Himalaya. Part II. *Journal of Bombay Natural History Society*, **50**, 559-572.

Wynter-Blyth, M.A. (1957). *Butterflies of the Indian region*. Today and Tomorrow's Printers and Publishers, New Delhi, p 523.

Yeragi, S.G. and Yeragi, S.S. (2006). Biodiversity and Present status of Avifauna in and around the Mangroves of Akshi creek, Alibaug. In: *Biodiversity and Environment* (Eds.: Pandey B.N. and Kulkarni G.K.). A.P.H. Publication, New Delhi, 29-34.

Index

A

B

C

D

E

F

G

H

I

K

L

M

N

O

P

R

S

T

U

V

X

Y

www.ingramcontent.com/pod-product-compliance
Ingram Content Group UK Ltd.
Pitfield, Milton Keynes, MK11 3LW, UK
UKHW021451280726
14060UKWH00001BA/349